CW00338864

CARDIFF INSTITUTE OF
LLANDAFF
HIGHER EDUCATION - LIBRARY

10126488

Monitoring Building Structures

Monitoring Building Structures

Edited by

J.F.A. MOORE
Head of Structural Engineering Branch
Building Regulations Division
Department of the Environment
London

and

formerly Head of Structural Integrity Division
Building Research Establishment
Watford

Blackie
Glasgow and London

Van Nostrand Reinhold
New York

Blackie and Son Ltd
Bishopbriggs, Glasgow G64 2NZ
and
7 Leicester Place, London WC2H 7BP

Published in the United States of America by
Van Nostrand Reinhold
115 Fifth Avenue
New York, New York 10003

Distributed in Canada by
Nelson Canada
1120 Birchmount Road
Scarborough, Ontario M1K 5G4, Canada

16 15 14 13 12 11 10 9 8 7 6 5 4 3 2 1

First published 1992

British Library Cataloguing in Publication Data

Monitoring building structures.
I. Moore, J.F.A.
690

ISBN 0-216-93141-X

In the USA and Canada

ISBN: 0-442-31333-0

Library of Congress Cataloging-in-Publication Data available

Phototypesetting by BPCC-AUP Glasgow Ltd
Printed in Great Britain by BPCC Wheatons Ltd, Exeter

Preface

There is an increasing number of buildings that require informed decisions to be made about their continued safety and serviceability. Although social and economic issues are often all-important influences, the technical issues nevertheless need to be addressed objectively, efficiently and reliably. This book shows how monitoring the physical behaviour of a structure can assist the engineer to meet these conditions when making an assessment.

The book is aimed primarily at the practising engineer charged with making recommendations in respect of safety and serviceability. By the same token, it will be of value to the client specifying a brief for assessment or evaluating the report of an investigation which involves monitoring. The book will also be one of reference for those engaged in research involving monitoring, and an aid to the advanced student who needs to understand better the full-scale performance in service of building structures.

The need to assess safety and serviceability may arise for a variety of reasons, ranging from problems developing in service to change of use or the introduction of innovative features at the design stage. These reasons are explored in the first chapter which establishes a philosophy by which the assessing engineer can determine appropriate courses of action. Observations and measurements which do not address the real issues are worthless but too much information which cannot be effectively digested and interpreted is also not useful. Subsequent chapters deal with different types of measurement, ranging from overall displacement to local strain or deformation, and from one-off measurements to installed schemes capable of operation over a significant period of time. The techniques therefore include conventional surveying, photogrammetry, automatic and autonomous systems and dynamic response, as the tools of principal practical value for monitoring building structures.

The description of each technique includes limits on its applicability, its general state of development and experience with use, precautions necessary in use to ensure reliable results, expected accuracy, and methods of interpretation and presentation of results. Indications are given of novel techniques and the prospects for their development and application to this field of measurement. Case histories are outlined for each type of approach and the book concludes with discussion of the process of developing a monitoring scheme.

I am greatly indebted to the contributors who have focused their expertise in the manner necessary for this book.

J.F.A.M.

Contributors

Mr Graham S.T. Armer Head of Structural Stability Section, Building Research Establishment, Garston, Watford

Dr Brian R. Ellis Head of Structural Dynamics Section, Building Research Establishment, Garston, Watford

Mr Alan Kenchington Structural Statics Ltd, Newberry House, Southgate Street, Winchester

Dr John F.A. Moore Building Regulations Division, Department of the Environment, 2 Marsham Street, London SW1P 3EB

Mr Brian J.R. Ping Director, Glen Surveys Ltd, 309 High Street, Orpington

Mr David Stevens AMC, Fordbrook Business Centre, Pewsey

Contents

1 Introduction

G.S.T. ARMER

1.1 Failure in the built environment

For most advanced economies, building construction represents one of the largest, if not the largest, single investment of national resources. The well-being of such economies is therefore heavily dependent upon the satisfactory performance of its construction. Failures in the built environment occur for a variety of reasons and on a variety of scales. In macroeconomic terms, the small domestic structure which burns to the ground because of a fire started in play by children is insignificant when compared with a national disaster such as a major earthquake, affecting a million people simultaneously. There are less obvious types of failure which can also cause widespread human distress and be of considerable economic significance, such as problems occurring in populations of system-built construction where many thousands of units have to be considered as unsatisfactory as a result of a few actual failures. Examples of this phenomenon of population failure have occurred in the UK in large-panel system-built housing (BRE, 1985; 1986; 1987) and in long-span school and leisure halls (Bate, 1974; 1984; DoE, 1974) amongst others.

In the more temperate climates, failure of the infrastructure usually causes relatively short-term, if expensive, problems. For example, the storms in 1987 and January, 1990, in the UK (Figure 1.1) caused disruption of road, rail, air and sea transport, the loss of electricity supply to around a million homes and overloading of the telephone system. For a short time following the latter storm, all the London railway termini were closed. These types of failure are most commonly associated with what may be called severe natural environmental conditions. It is unusual to experience widespread structural failure in these situations although, of course, there are often isolated cases which do occur.

In many areas of the world, such as southern Europe, parts of Australasia and the Americas, the environment can be much harsher with the potential for earthquakes, hurricanes or typhoons to cause substantial destruction with consequent high costs, long repair or replacement times, and great costs in terms of human distress (Figure 1.2).

The design of structural systems to resist the forces generated in each of these extreme load conditions differs between the two cases. However, in both

Figure 1.1 Wind damage to an industrial steel-framed building.

instances the designer is expected to create a built environment which achieves a level of performance acceptable to the public. That is, acceptable in terms of the perceived severity of the hazard. In the event of a major earthquake or typhoon, when the failure of domestic and commercial buildings would not be surprising, then essential construction such as hospitals, power stations and emergency services accommodation would be expected to survive (Figure 1.3).

1.2 Structural design philosophy

The design options available to the structural engineer to meet his clients' and the public's requirements for buildings which perform satisfactorily within various hazard scenarios may be categorised (Armer, 1988) quite simply thus:

(a) Provide a strong basic structure.
(b) Provide redundancy within the basic structure.
(c) Provide sacrificial defence for the structure.

Figure 1.2 Storm damage in Australia.

Figure 1.3 Structural damage due to an earthquake.

(d) Provide monitoring systems which warn of inadequate structural conditions.

The first three of these are relatively familiar ground for the designer.

The principle of making load-bearing structures strong enough to withstand all the loads which are likely to impinge during their lifetime is central to the universally applied structural design philosophy which has developed since the eighteenth century. It provides a straightforward basis on which to establish the level of structural safety by admitting the use of factors attached to either the design loads or to the response, i.e. the strength descriptor. By so doing, the designer is given a direct indication of the margins of safety against failure under the loads he has explicitly identified. The success of this approach depends heavily upon the engineer's skills in predicting both the possible ranges and the combinations of loads which the structure will have to carry and the material and structural performance of the building itself.

The principle of explicitly incorporating redundancy into a structure has more recent antecedents. In the early days of aircraft design, there was a major problem with the provision of sufficient power to get the machines off the ground. The two obvious strategies to surmount this difficulty were to build more powerful engines and/or to reduce the weight of the aeroplanes. The latter approach led to some very elegant mathematical solutions for the problem of 'minimum weight design', which in turn led to brittle behaviour of the resultant airframe structures! To overcome this particular difficulty, the concept of 'fail-safe' design was developed, and the idea of beneficial redundancy was introduced into the designer's vocabulary. By the intelligent use of redundant elements, it is possible to ensure that the failure of a single element does not precipitate the failure of the complete structure. As yet, there are no satisfactory theories to guide the designer in this field, but experience and empiricism must suffice.

The provision of sacrificial defence for a building structure is even less scientifically founded. The use of bollards and crash barriers to prevent vehicle impacts is commonplace and their efficacy proven in practice. The use of venting elements such as windows and weak structural elements to reduce the level of explosive pressure on major structural elements is less widely adopted. Weighted trap-doors are sometimes used in factory buildings where severe cloud explosions are a high risk, but are arguably not sacrificial. Mainstone (1971) has given considerable data for the design of explosion venting windows, but great care must obviously be taken when using this design strategy. The provision of in-service structural monitoring systems in building construction has so far been limited to a few isolated cases such as the commercial fair complex in the USA (IABSE, 1987). Some systems have found application in civil engineering construction, for example in special bridges and nuclear power stations. It is probably reasonable to assume that the use of monitoring in these circumstances reflects a 'belt and braces' approach to safety rather than a planned use of the technique in an integrated design philosophy.

1.3 The role of monitoring

A dictionary definition for 'to monitor' is: 'to watch or listen to (something) carefully over a certain period of time for a special purpose'. The end of this definition is perhaps the most important part since it identifies the need to establish purpose as an essential element of the activity of monitoring. There have been a number of examples where considerable quantities of data have been collected, particularly in the field of research, in the vain hope that some brilliant thought will arise as to what should be done with this information. Many who have been associated with structural testing will be familiar with the thesis that putting on a few more gauges would be 'useful' since more information must be helpful. Unfortunately, insight rarely arrives in such circumstances and, if it does, the inevitable conclusion is that something else should have been measured! So before any consideration is given to methods of observing the performance of construction, the question must be asked: 'why is this expensive process to be established?' The answer to this question necessarily involves clear understanding of the positive actions which could result from the data-gathering exercise.

The role of construction monitoring has to be established against a background of requirements emanating from the public (i.e. the state), the owner and the user. These requirements have to be expressed in a form which is compatible with the data generated by any monitoring systems which are established. Putting this point more explicitly, whilst it is relatively easy to gather enormous amounts of data from most monitoring systems, it is usually very difficult to interpret such data and to develop useful consequential responses.

Public concern with the performance of building structures is principally directed at those aspects which impinge on the individual, that is upon his own safety and welfare. The acceptable degree of safety is encapsulated in the various building regulations, standards and codes of practice. Unfortunately, there is not a simple correlation between the safety of people in and around construction and the security/safety of that construction, as shown by Armer (1988). Neither is it possible to express a level of safety for a construction as a number. In spite of the large number of academic papers which purport to discuss safety levels in explicit terms and to offer the designer the opportunity to choose his own value, the concept is entirely subjective and acceptability will vary with construction type and location, current political situation and so on.

Any post-construction monitoring of building works carried out by government or state authorities acting as the public executive rather than the owner is essentially limited to the collection of statistical data on failures. This form of monitoring matches the nature of the response which can be implemented practically by a national regulating body. For example, certain types of construction can be outlawed or special requirements can be included in regulations, such as the provision for buildings of over four storeys to be designed to resist disproportionate collapse following accidental damage. The effect of this form of monitoring and response is the control of the performance of the population of constructions and *not* that of an individual building. This point is sometimes misunderstood in discussions on the function of codes of practice for good design. By giving rules for the design of construction, i.e. codes of practice, the regulating body can ensure that a particular building is part of the population of buildings modelled by a particular code, but it does not, however, guarantee further its quality. Since within any population of artefacts the performance of individual elements will vary from good through average to bad, such will be the lot of construction. Thus the objective of monitoring construction by public authorities is twofold: firstly, to ensure that acceptable levels of safety are sustained for the people and, secondly, to ensure that a degree of consumer protection is provided to meet the same end.

The owner of a building, if not also the user, will require to protect his investment both as capital and as a generator of funds. This can be achieved by regular monitoring and by consequential appropriate maintenance. There are, of course, many quirks in the financial world which may make this simplistic description somewhat inaccurate for particular owners and at particular times but nevertheless it represents a realistic generality.

The user will be concerned that the building he rents or leases provides a safe working environment for both his staff and his business. It is therefore in the user's interest to monitor the building he occupies to ensure that his business is not damaged by the loss of facility.

It is therefore of concern at many levels to ensure the proper functioning of the building stock, and a necessary part of this process is to monitor current condition.

The discussion so far is predicated on the thesis that monitoring construction should be a legitimate weapon in the armoury of those concerned with the in-service life of buildings. It must be admitted, however, that its use in the roles described is only in its infancy. Most practical experience in the instrumentation of building has been gained by researchers. Their objectives have usually been to validate, or sometimes to calibrate, theoretical models of structural behaviour. By so doing they hope to provide another design aid for the engineer. In the technical literature, there are many reported examples of this use of the technique. Ellis and Littler (1988) on the dynamics of tall buildings, Sanada *et al.* (1982) on chimneys and Jolly and Moy (1987) on a factory building illustrate the variety of work undertaken so far.

Since there is quite a limited number of structural performance indicators which can be monitored, and likewise a limited number of suitable (which is really a euphemism for stable) instruments, there is considerable scope for the exchange of techniques and methods between the various applications for which some experience exists.

References

Armer, G.S.T. (1988) Structural safety: Some problems of Achievement and Control. *The Second Century of the Skyscraper*. Van Nostrand Reinhold, New York.

Bate, S.C.C. (1974) Report on the failure of roof beams at Sir John Cass's Foundation and Red Coat Church of England Secondary School. *BRE Current Paper CP58/74*, Building Research Establishment, Watford.

Bate, S.C.C. (1984) High alumina cement concrete in existing building superstructures. *BRE Report*. HMSO, London.

BRE (1985) The structure of Ronan Point and other Taylor Woodrow-Anglian buildings. *BRE Report BR 63*. Building Research Establishment, Watford.

BRE (1986) Large panel system dwellings: Preliminary information on ownership and condition. *BRE Report BR 74*, Building Research Establishment, Watford.

BRE (1987) The structural adequacy and durability of large panel system buildings. *BRE Report BR 107*. Building Research Establishment, Watford.

DoE (1974) Collapse of roof beams. *Circular Letter BRA/1086/2* (May 1974). Department of the Environment, London.

Ellis, B.R. and Littler, J.D. (1988) Dynamic response of nine similar tower blocks, *Journal of Wind Engineering and Industrial Aerodynamics*, **28**, 339–349.

IABSE (1987) Monitoring of large scale structures and assessment of their safety. *Proceedings of Conference, Bergamo, Italy*.

Jolly, C.K. and Moy, S.S.J. (1987) Instrumentation and Monitoring of a Factory Building. *Structural Assessment*. Butterworths, London.

Mainstone, R.J. (1971) The breakage of glass windows by gas explosions. *BRE Current Paper CP 26/71*. Building Research Establishment, Watford.

Sanada, S., Nakamura, E.A., Yoshida, M. *et al.* (1982) Full-scale measurement of excitation and wind force on a 200 m concrete chimney. *Proceedings of the 7th Symposium on Wind Engineering, Tokyo*.

2 Surveying

J.F.A. MOORE and B.J.R. PING

2.1 Introduction

2.1.1 Scope

Surveying is used here as a term to cover a wide range of techniques which rely primarily on visual observation with instrumentation to obtain a measurement of length or determination of position from which movement may be deduced. The methods generally require the presence of an observer on site at the time of measurement, as opposed to the techniques described in chapter 5, which can be operated remotely once installed.

This does not mean that the techniques are necessarily unsophisticated or that they may not employ or require advanced computational facilities, but rather that there is a wide range from which to select the system most appropriate in terms of cost and accuracy to study effectively the problem in hand, as emphasised in chapter 1.

2.1.2 Accuracy

Many of the principles of these techniques are common to those used in traditional site-surveying techniques familiar to engineers setting out buildings prior to and during construction or by land surveyors at the planning stage. The additional characteristics required for monitoring the performance of a building structure include the ability to refer reproducibly to points on the structure over significant periods of time, with confidence in the integrity of the points. There is the corresponding need for reliable stable datum points, if absolute, rather than just relative, measures of movement are required. The requirements for accuracy of monitoring are likely to be considerably greater and therefore more onerous to achieve.

The movements expected during a monitoring programme will depend on the size and type of the building and the cause of movement. The effects of foundation movement on superstructure, dimensional changes caused by environmental effects, structural deflection under load or incipient instability may be of quite different magnitudes and patterns. Nevertheless, the expected magnitude is likely to be millimetres, up to perhaps 10 or 20, but of course with exceptions, and the accuracy required to be no worse than a millimetre. If progress of the movement is to be followed, particularly if it is to be used to trigger action to safeguard safety as opposed to leading to an understanding of the behaviour of the building structure or fabric, or if it is required to determine when movement is ceasing, then submillimetric accuracy may be required.

The factors to be considered in order to achieve the required accuracy may be enumerated after Cheney (1980) as:

(1) The accuracy of the basic survey instruments used.
(2) The precision of any necessary ancillary equipment.
(3) The rigidity of fixing and long-term integrity of permanent reference fittings to which measurements are to be made.
(4) The repeatability of positioning of all equipment used.
(5) The stability of points taken as data.
(6) The relative position of points to be monitored on the building and of the datum points.
(7) The effects of meteorological conditions.
(8) The competence of the observers.

Compatibility of all these factors is essential for the attainment of accurate determination of position at one time and for the reliable comparison of observations made at different times, often at long intervals.

When a survey is to be designed to monitor the response to known events, such as adjacent construction work, it is essential that sufficient time is allowed to permit at least two sets of observations to establish datum measurements before possible disturbance takes place.

2.1.3 Types of measurement

The remaining sections of this chapter identify the principal approaches to each type of measurement and the main issues to be considered in deciding on a scheme. It will also be necessary to decide in advance the criteria against which to judge the results; they are particular to the individual building and beyond the scope of this book.

Determination of vertical movement is probably the most common requirement, whether for a complete structure, e.g. differential settlement, or for a component, e.g. floor deflection, and conventional precise survey techniques are widely used. Systems based on electrical transducers are discussed in chapter 5 and dynamic behaviour such as vibration of floors in chapter 4.

Out of plumb may be simply a reflection of the building as constructed or an indication of response to, for example, loss of restraint caused by corrosion of metal tying or an adjacent excavation. Very simple techniques may be suitable if the accuracy attainable with precise optical plummets is not required; again, automated laser systems mentioned in chapter 5 may be appropriate.

Measurement of horizontal displacement in general will require precise triangulation or trilateration, intersection or bearing, and distance measurement, mechanical or electronic. If the direction of movement can be assessed with certainty then a variety of direct or remote reading devices (chapter 5) may be fabricated to fit the circumstances.

Cracks are likely to be the most common reason for deciding in the first place to monitor a building in a more comprehensive way. They may occur for a variety of reasons in various construction materials and their significance may range from structural inadequacy to weathertightness and appearance. Sometimes their importance may be determined from a survey of 'present condition', but more often assessment of any continuing movement or an understanding of the nature of the movement will require measurements of change in crack width over sufficiently long periods of time.

2.2 Vertical position and movement

2.2.1 Types of approach

The techniques considered here are optical levelling for measuring position with time, and thus vertical movement, and simple liquid levelling to measure existing differential position. Precise liquid levelling to determine vertical movement is considered in chapter 5.

2.2.2 Optical levelling equipment

Unless it is obvious that large movements are occurring and it is clear that use of a 'surveyor's level and staff' yielding an accuracy of a millimetre or so will be adequate, it will be necessary to use a geodetic level with optical micrometer in conjunction with an invar staff to obtain submillimetric accuracy. The accuracy of automatic levelling instruments is such that their use has almost replaced that of the best split-bubble or 'tilting-head' instruments, so improving the reliability and speed of use in the field. There are over a dozen manufacturers worldwide, and their recommendations for checking and maintenance should be followed by the engineer doing the measurement, unless this task and associated responsibilities are contracted to a professional surveyor.

The accuracy achievable will depend on the number of legs before a round can be closed on the starting point and the ability to limit and equalise the length of sights and to maintain precise collimation. Cheney (1980) quotes closing errors of 0.3 mm over 10 change points with a maximum limit of sight of about 15 m.

Careful and systematic procedures are required by the operator for making observations, recording them and processing the results. Cheney (1973) has described this in detail.

2.2.3 *Datum points*

When absolute measurements of movement are required it is essential to establish one or more reliable stable datum points to which each successive set of observations may be made. If a building is sufficiently massive and well-founded, a datum point of the type described in section 2.2.4 should be satisfactory. In built-up areas, especially where there is or has been much underground construction, or in soil conditions particularly subject to moisture effects, it may be necessary to select positions some distance or some depth from the building to be monitored.

However, further errors may be introduced in carrying the levels over increased distances from the datum, so a balance has to be struck. The difficulties may be overcome by using temporary intermediate bench marks (TBM) and checking differences between adjacent TBMs to assess their stability. This procedure has been used satisfactorily over distances of 1 km.

The design of a sleeved datum or bench mark founded some 6 m below the surface is described in detail by Cheney (1973), and alternative concepts which may be used also for establishing horizontal movement (see section 2.4 below) are described by Burland and Moore (1973). The subsurface reference point described by them has been used satisfactorily at many places in the streets of London for monitoring the effects of neighbouring construction on existing buildings, where it is impractical to provide a permanent reference point above the surface and where it is necessary to achieve stability on a firm stratum away from surface effects. It is particularly important that clients appreciate that big city buildings are not necessarily stable and that the cost of establishing reliable datum points will be money well spent.

2.2.4 *Levelling stations*

Confidence in the assured performance of levelling stations suggests that they should be purpose-made. In order that their prime function of accurately accepting a levelling staff can be met they should have other attributes. They should be capable of easy, reliable installation, they should be vandal-proof and resistant to deterioration, preferably for many years, and they should be unobtrusive and visually acceptable.

The BRE levelling station (Cheney 1973) is used widely and consists of a stainless-steel socket 65 mm long and 22 mm in diameter which is fixed to the structural component to be observed. It is sealed when not in use by a plastic bung with its outer surface flush with the surrounding finish, shown in section in Figure 2.1. A removable stainless-steel plug, threaded at the inner end, is located to a radial accuracy of 0.03 mm by a spigot and socket. The other end of the plug is spherical to accept the foot of a levelling staff (Figure 2.2). Normally only one plug would be used for a site, with a second in reserve. The same concept for achieving positional repeatability can be

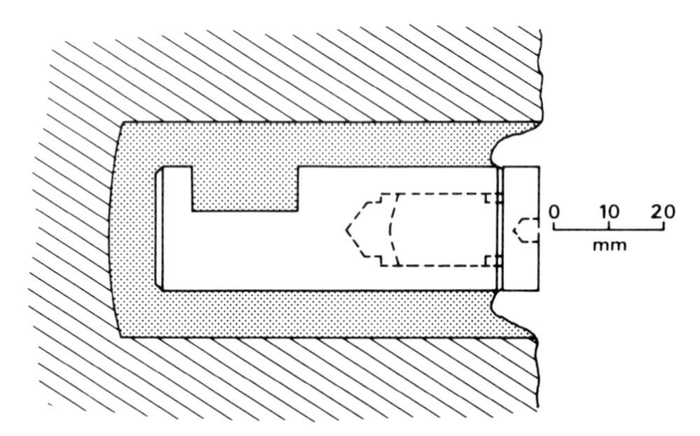

Figure 2.1 Cross-section of a Building Research Establishment (BRE) levelling station.

Figure 2.2 Staff on BRE levelling station.

Figure 2.3 Taping to a BRE levelling station.

extended to other attachments to this type of reference point, e.g. taping (Figure 2.3) as part of a bearing and distance system as in section 2.4.4.

2.2.5 Procedure and results

Detailed discussion of the practical steps necessary to use precise levelling equipment to its full potential in the context of a building and a suggested procedure for presentation of the observations so as to draw attention to potential sources of error are given by Cheney (1973) (Figure 2.4). Figure 2.5 illustrates how complete performance may be displayed in relation to possible causal events.

Plots of settlement of the four corners of a building readily draw attention to tilt and twist (Figure 2.6) of a house constructed on made-up ground in an old excavation.

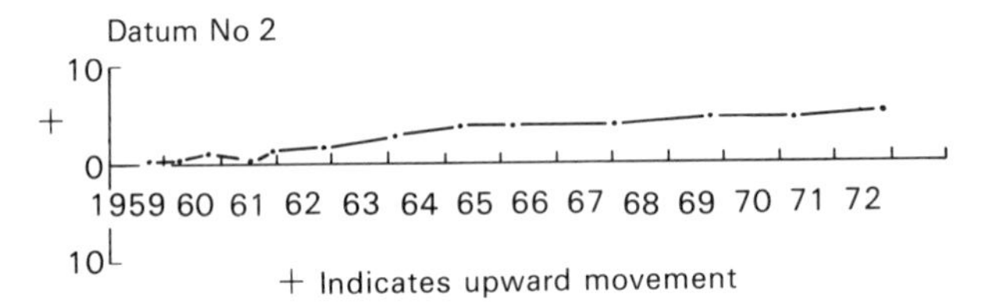

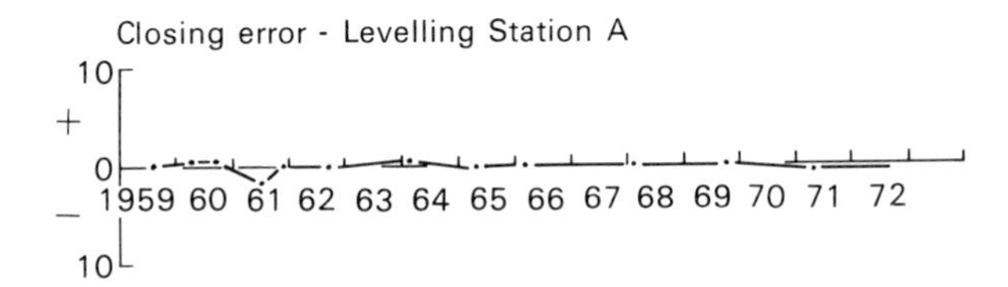

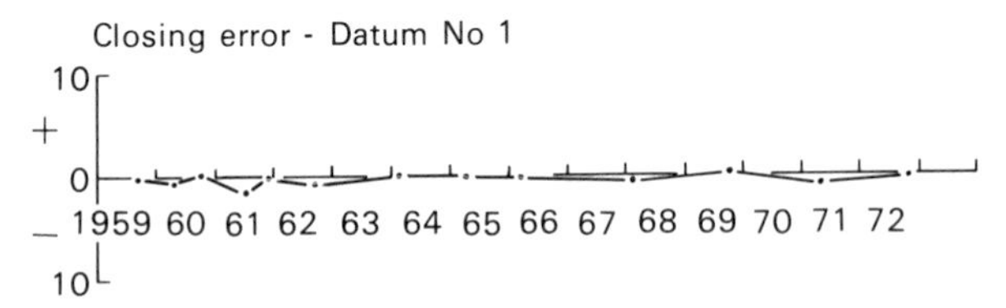

Figure 2.4 Example of levelling showing closing errors. After Cheney (1973).

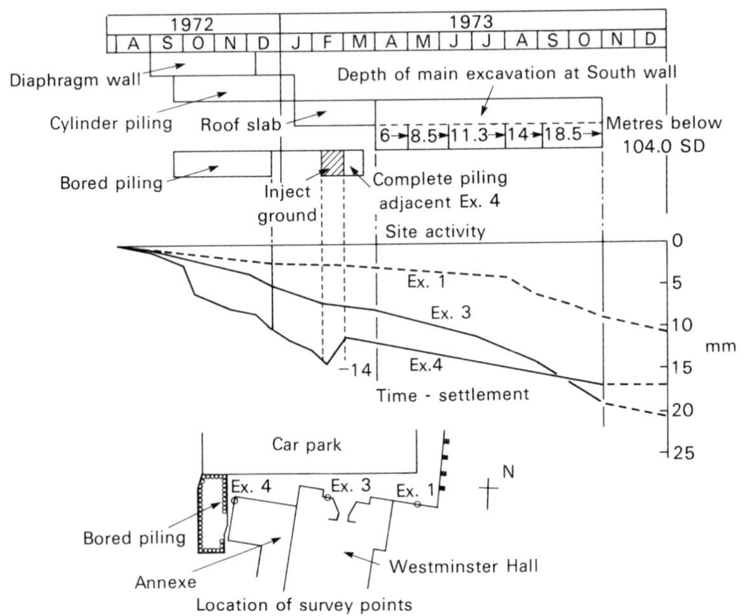

Figure 2.5. Settlement related to adjacent construction activity. After Burland and Hancock (1977).

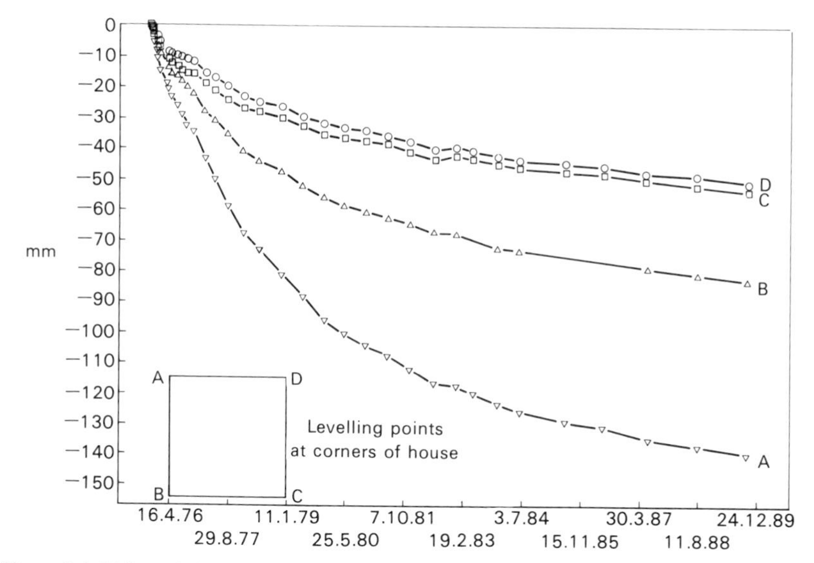

Figure 2.6 Differential settlement of a house on made-up ground, with thanks to D. Burford.

However, developments in automatic instruments have been matched with the use of computerised data capture at the time of observation, e.g. hand-held loggers and on-site computation. Errors may be assessed and distributed automatically to yield present levels and changes from previous observations, as in Table 2.1.

2.2.6 Simple liquid levelling

This method is particularly useful for measuring existing differential settlement or heave in masonry buildings on the assumption that the mortar bed-joints or damp-proof course were built nominally level. The accuracy achievable in normal building construction thus limits the accuracy of this method to about ± 2 mm, although observations may be recorded to the nearest 1 mm.

A suitable device comprises a reservoir positioned so that the liquid in it is about half a metre above the course of masonry to be followed along the building. A flexible transparent tube is attached at one end near the bottom of the reservoir and at the other to a scale which acts as a levelling staff. The water level is then read against the staff, first against a chosen datum and then at successive positions along the chosen course. A full description of the procedure for setting up and use is given by BRE (1989a), where it is noted that the technique may be used also for measuring the 'levelness' of floors.

Table 2.1 Example of modern level print-out after error correction. Docklands Light Railway—surface and building monitoring task: Tower of London Moat Wall—ground levelling

Survey no. 49 Date 15/11/89 Sheet 10

Point ID	Survey no. 26	Survey no. 48	Survey no. 49	Overall difference	Latest difference	Remarks
ST1	11.860	11.838	11.834	−26	−4	
ST2	11.796	11.775	11.769	−27	−6	
ST3	11.783	11.762	11.758	−26	−4	
ST4	11.766	11.745	11.740	−25	−4	
ST5	11.765	11.743	11.739	−26	−4	
ST6	11.767	11.745	11.741	−26	−4	
ST7	11.767	11.746	11.740	−27	−6	
ST8	11.781	11.759	11.755	−26	−3	
ST9	11.775	11.753	11.749	−26	−4	
ST10	11.776	11.754	11.750	−26	−3	
ST11	11.761	11.739	11.735	−25	−3	
ST12	11.740	11.718	11.715	−25	−3	
ST13/N	11.704	11.683	11.680	−24	−3	
ST14	11.686	11.664	11.660	−26	−4	
ST15	11.679	11.660	11.656	−23	−3	
ST16	11.668	11.650	11.646	−22	−4	
ST17	11.597	11.582	11.575	−22	−7	
ST18	11.516	11.500	11.493	−22	−6	
ST19	11.446	11.430	11.426	−20	−5	
ST20	11.370	11.356	11.352	−19	−4	
ST21	11.349	11.336	11.332	−17	−4	
ST22	11.303	11.290	11.287	−16	−3	
ST23	11.271	11.259				Destroyed
ST24	11.240	11.230	11.227	−13	−3	
ST25	11.227	11.217	11.215	−13	−3	
ST26	11.201	11.193	11.191	−10	−2	
ST27	11.216					Destroyed
ST28	11.244	11.237	11.234	−10	−3	
ST29	11.282	11.275	11.273	−9	−2	
ST30						
ST31	11.381	11.376	11.372	−8	−4	
ST32	11.428	11.423	11.419	−8	−4	
ST33	11.498	11.493	11.489	−9	−4	
ST34	11.581					
ST35	11.632	11.630	11.627	−5	−3	
ST36	11.687	11.685	11.682	−5	−3	
ST37	11.745	11.743	11.740	−5	−3	
ST38/N	11.888	11.884	11.881	−6	−3	

2.3 Plumb

Measuring 'out of plumb' of a wall may be achieved using a theodolite of 10-second accuracy if sufficient access beyond and in line with the wall is available. However, on a low-rise building sufficient accuracy may be achieved using a plumb bob and scale if certain elementary precautions are taken and procedures followed. BRE (1989a) describes a process which assumes that a two-storey building was not out-of-plumb by more than ±20 mm when built and enables measurements to be made to ±1 mm. It relies on stabilisation of the bob on the end of a line by immersion in oil and measuring the offset of the wall from the line using a ruler.

The same technique of a damped pendulum may be used over much greater heights—six storeys or more. Great care is required to avoid disturbing the pendulum once stabilised, and if observations are to be made at several heights a remotely operated continuous system using electrical transducers of the general type discussed in chapter 5 would be more appropriate.

If change of plumb is to be monitored, particularly the top of a tower or lift shaft relative to the base, an optical plummet is required. This instrument enables a vertical line of sight to be established. A reference pillar should be established at the base on which the instrument can be mounted when observations are required. Normally an instrument with an accuracy of 1 second is satisfactory and equivalent autoplumbs are available to facilitate the operation without loss of accuracy. A suitable sighting target should be established at the upper level or levels and the angular tilt from vertical required to sight the target observed at datum and subsequent times. Measurements in orthogonal directions enable the components of horizontal displacement of the target relative to the base to be calculated.

An installation in a clock tower some 100 m high required the upper target to be removable. The sighting target was attached to a beam spanning the stairwell at a height of about 55 m and supported on kinematic supports. These comprised at one end a three-point location consisting of an equilateral triangle of three 5-mm balls seated in a cone, a V-groove and a flat, and at the other a single ball on a flat. This system ensured that the target was replaced after removal in the same position relative to the support at one end. The results of tracking the movement over a period of more than $2\frac{1}{2}$ years are shown in Figure 2.7 from Burland and Hancock (1977). Correlation

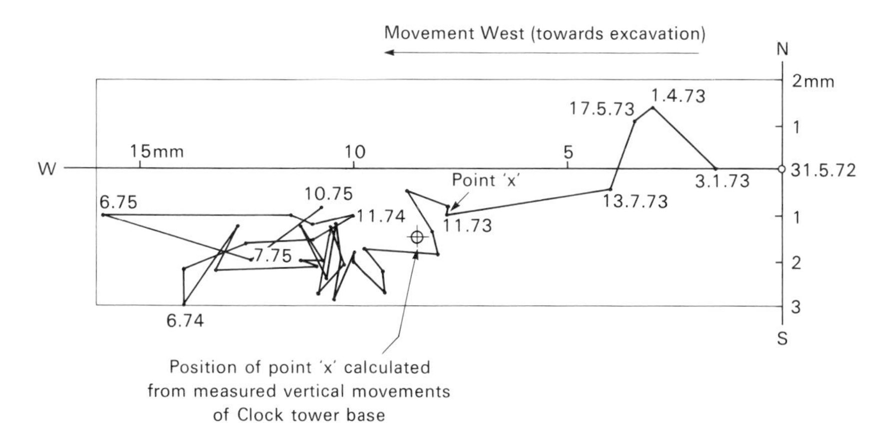

Figure 2.7 Lateral movement of the clock tower. After Burland and Hancock (1977).

with an alternative measurement of the displacement at height, deduced from levelling the base of the tower and assuming that it tilted as a rigid body, is given by point 'x' in Figure 2.7.

The same effect may be achieved now using a laser to provide the 'line of sight', the position of the light beam being read visually or sensed opto-electrically on a calibrated target. Targets catering for 100 mm movement may detect position to an accuracy of ± 0.1 mm, although over a range of about 100 m the accuracy may be ± 0.2 mm.

The accuracy with both approaches depends on the accuracy of the plummet or the long-term stability of the laser, on problems of refraction in a long column of air and, of course, on the distance of the target from the instrument.

2.4 Horizontal position and movement

2.4.1 Approaches

Over the last 20 years there have been rapid improvements in the equipment available for measuring angles and distances. Electronic theodolites and distance meters are commonplace. The surveyor's traditional tool for measuring accurate height differences (see section 2.2), a quality level with parallel plate micrometer and invar staff, has seen very little change or improvement. Hence it is still simpler and many times cheaper to derive submillimetre accuracy in height than it is in plan, for a large array of scattered points.

Earlier sections described how horizontal movement can be detected in one horizontal plane by simple devices such as plumb bobs, tell-tales, lasers or autoplumbs. However, when relative or absolute movements are required of a series of points on different planes then measurement can become very time-consuming, labour-intensive and costly.

There are four basic concepts of trigonometry available to derive coordinates; triangulation, trilateration, intersection or bearing and distance. The first two are seldom used because of the requirement to occupy the point to be monitored. Intersection can include either a pair or more distances or angles. Equally bearings and distances can be read from more than one position, which provides several solutions, including angular or distance intersection.

As triangulation and trilateration necessitate occupying the points to be monitored, the monitoring descriptions are subdivided hereafter into the general headings: occupied points and remote points. The requirements for at least two stable reference points are similar to those described in section 2.2.3.

2.4.2 Occupied points

2.4.2.1 Triangulation. Triangulation, the surveyor's original concept for accurate control before it became possible to measure long distances quickly by electronic distance meters, still has an application in monitoring. In triangulation all angles in a triangle are measured and therefore the principle is only applicable when the points to be monitored can be occupied. In most monitoring situations the points to be monitored are inaccessible or in awkward positions for setting up tripods, hence remote measuring techniques such as intersection or bearing and distance are used.

A good example of using triangulation for monitoring is given by a river defence trial which, although some 10 years old, emphasises the point about accessibility. On this project engineers needed to know the movement of several passive monitoring points and also active instruments such as piezometers and inclinometers. The river defence trial lasted several months and had to include all tide cycles. On this occasion a Wild T3 theodolite was used, measuring to 0.1 second of arc. Scale was provided by a single baseline measured using a Mekometer to 0.1 mm accuracy.

The measuring exercise was very slow and time-consuming, requiring several rounds of angles to each point. Reduction of angles and triangulation computing was also a slow and laborious process. With modern electronic theodolites with dual-axis compensation, measurement to the same accuracy can be achieved with half the number of observations, as it is unnecessary to observe on two faces. Angles can be logged automatically and the data can be passed directly to computer for calculation.

Triangulation is still capable of giving the accuracy required for monitoring, but in terms of speed of data capture and ability to derive coordinates quickly this technique has been overtaken by 'real-time' solutions, which are addressed later.

Triangulation is only appropriate if the monitoring point array provides triangles of good cut—ideally no angles below 30° or greater than 120°.

2.4.2.2 Trilateration. Trilateration is the equivalent principle to triangulation in which all sides of triangles have to be measured as opposed to all three angles. With modern instrumentation such as the Mekometer it is possible to measure distances to 0.1 mm (±1.5–2 parts per million) and therefore high accuracy can be achieved using this principle. The Mekometer itself is slow, taking some 2 minutes to derive a distance. There are, however, at least two competitors of the Mekometer, the Geomensor and the Tellumatt 200, which are much quicker, of similar precision and significantly cheaper.

As with triangulation, trilateration can only be used where all monitoring points can be occupied. It is certainly quicker than measuring rounds of angles and hence would be applicable to the river trial bank application. However, the most likely application of trilateration would be in the civil engineering field, say over a large dam site where range and/or heat haze might make it difficult to observe angles. Although the sight path of both theodolites and electronic distance measurements (EDM) may be affected by atmospheric conditions, corrections with EDM may be made automatically and the pointing at the target sensed electronically as opposed to the optical bisection required using a theodolite.

Angle of cut of the distance rays is not so critical as it is for triangulation. The method produces fewer data than triangulation but it is still necessary to use a computer to derive a solution. Least-squares analysis of the interlinking net of information is imperative.

Generally, therefore, the most widely used monitoring techniques are intersection and bearing and distance. They are most appropriate for a large array of points, over relatively short ranges. Using these methods control points have to be established and constantly remeasured to high accuracy to derive relationships for the monitoring points.

However, triangulation and trilateration, by their very principles, interrelate all points to each other, and therefore it is often possible to incorporate the control, i.e. datum points, into the trigonometrical framework of monitoring points. This is not the case for intersection or bearing and distance, hence a monitoring project may involve a combination of techniques. If control is not provided for a remote point scheme by traversing, then either triangulation or trilateration would have to be employed.

2.4.3 Remote points

2.4.3.1 Principle. Intersection or bearing and distance are the only principles that a surveyor can apply using theodolites, with tapes or distance meters, if the points to be monitored cannot be occupied.

If targets cannot be placed on the structure to be monitored then the principle of intersection is the most applicable solution. However, there is presently one distance meter, the Wild DIOR 3002, that can derive distances without the use of retroreflectors at ranges up to 250 metres. The resolution of this instrument is ±5 mm, and as one is usually attempting to measure to at least millmetric accuracy this instrument would seldom be adequate. Alternatively, the bearing and distance technique could be used, if it was possible to tape distances to a satisfactory accuracy.

2.4.3.2 Intersection. Traditional 1-second theodolites can still be used to provide quality information, dependent on the accuracy required, the range from the points to be monitored, the quantity of points to be monitored and the time available.

An angle read to a quality of ±3 seconds subtends 0.5 mm at 34.3 metres. A pair of rays, of 3-second quality, can therefore be assumed to provide an intersection of ±1 mm accuracy. Hence at close range optical mechanical theodolite angle intersection is viable. There are, however, several other factors which affect the quality of results whatever the instrumentation and methodology used, which are discussed in section 2.4.6. The most limiting factor of using traditional theodolites is speed of operation and volume of data to observe and process. To achieve 3-second accuracy with a 1-second theodolite, it is normal practice to observe two rounds of angles to each monitoring point on two faces, i.e. four pointings to the targets. This would have to be repeated from a second theodolite position to provide an intersection and hence a total of eight pointings per monitoring target, plus pointings to a reference object.

With expense the time/quantity problem of having to observe so many points can be overcome. There are now available electronic 'real-time' systems. Such systems employ electronic theodolites linked to a computer. As stated, the use of dual-axis compensated electronic theodolites eliminates the need to take readings on both faces of the circles. The use of the computer to give results which can be checked on-site eliminates the need to take more than one round of intersection angles as a separate check. Hence the traditional four pointings from each control point are reduced to one with an enormous saving in observation time. The computer brings many advantages to the monitoring process, apart from the obvious benefit of 'real-time' coordinates. These benefits eliminate hours of post-observation processing and the uncertainty as to whether the data are reliable. Reliability can only be determined from the computations. Further, because of the ability to store previous data, the linked theodolite computer system uses the repetitiveness of monitoring to the following special advantages:

(1) The computer can compare previous coordinates with present values so that gross errors of two theodolites observing two different targets can be eliminated. Observing the wrong point is always possible when one has a large array of remote targets. Spatially, two angles will always give a solution, but this solution will be way outside the predicted movement tolerance, and hence dissimilar pointings can be eliminated.
(2) Because data capture is so fast it is practicable to read all the intersected targets from at least one further control point. The computer can cope with an extra ray to each target and derive two further intersection solutions. These can be compared at the time of observation with the first set of coordinates. Differences in sets of coordinates, beyond acceptable tolerances, would indicate observation errors and alert one to remeasure. Hence, there should be absolutely no reason why data errors are found after site demobilisation.
(3) The computer comparison referred to in (1) above also determines the apparent movement for each point between sequential surveys. Small differences outside the predicted movements would alert the surveyor to take further checks. Using traditional equipment and processing off-site, anomalies that appear beyond predicted movements necessitate return visits to the site. Such return visits could hold up a construction project if the anomalies are alarming.

These three benefits of on-site 'real-time' computer comparisons can offset the high capital cost of the equipment. At the present time a high-specification monitoring system of two electronic theodolites coupled to a computer with prescribed software costs a minimum of £50 000.

A system such as this is presently being used on a 5-year monitoring programme at the Mansion House in London. With tunnelling for Docklands Light Railway taking place underneath, 'real-time' coordinate comparisons are vital to determine whether tunnelling should be halted because of either very rapid structural movement or differential movement which could lead to serious damage to the building.

The Mansion House scheme could have been carried out, probably more cheaply, by the bearing and distance technique described in the next section. The 'real-time' intersection requires two and sometimes three instruments observing simultaneously, and hence two or three observers. The bearing and distance solution employing a modern total station also provides a 'real-time' solution using just one instrument and one operator. It also requires fewer control points and less complex control geometry because there is no requirement to see the targets from several different directions. It was decided to adopt the intersection method because of targeting considerations, which are referred to in section 2.4.4.

2.4.3.3 Bearing and distance. A total of 250 points were monitored recently on the Moat Wall at the Tower of London, which was potentially affected by tunnelling for Docklands Light Railway. During tunnelling all 250 points were required to be observed, computed, compared with previous data sets every 24 hours and presented to the client for consideration as to whether tunnelling could proceed or not. This operation, using 1-second theodolites, would have required some 2500 pointings per day (including reference object checks) and up to 10 surveyors. Two hundred and fifty pointings per surveyor might seem low, but all angles would have to be reduced on-site and any anomalies remeasured instantly. On-site a computer would have been required to derive intersection coordinates, compare results with previous data sets and tabulate results for the client.

The solution to this mammoth manpower and processing problem was a modern total station (electronic theodolite and integral electronic distance meter). Only 250 pointings were necessary, no angle reductions were necessary in the field, and all data could be passed directly to the office computer by electronic data logger. In fact duplicate bearing and distance observations from different survey control points were observed. This provided up to four possible coordinate solutions per point (two individual bearings and distances and intersection by both angle and distance). This volume of redundant information allowed an assessment of the surveying accuracies and identification of surveying errors or anomalies which did not show up from field comparisons of present, 'real-time' coordinates and previous values.

Initial comparison was a visual inspection of the instrument-displayed coordinates and a printed list of previous values. Modern technology saved time and money, but did not enhance the quality of the results. This survey was carried out using a top-of-the-range total station with a specification of 2 seconds of arc and 2 mm + 2 ppm for distance. Price would be of the order of £14 000. Dual-axis compensation minimised most standard angular errors and collimation correction was applied, hence permitting single pointings. Distances were enhanced by applying cyclic error factor adjustments obtained by calibration of the total station against an interferometer.

Presently there is no subsecond, submillimetre total station in production. For higher accuracy there are electronic theodolites resolving to 0.3 second of arc and independent distance meters measuring to 0.1 mm. These have to be used separately. A pair of such instruments could cost £70 000 and are not readily available for hire. Using separate instruments for angle and distance would have doubled the measuring time. The intersection technique employed at the Mansion House would clearly have doubled the cost of the exercise at the Tower of London because manpower and equipment resources would have been doubled. However, the costs of manufacturing and installing targets at the Tower of London was many times more expensive than at the Mansion House. Hence in choosing the appropriate system for a particular job one cannot just consider instrumentation and manpower alone (see section 2.4.4).

Modern electronic instrumentation has revolutionised the speed of data capture and in some instances the ease of achieving high accuracy. However, the availability of such hardware can introduce conflict. Those requiring monitoring information often expect work to be carried out more quickly and more cheaply and tend to overspecify accuracy requirements. Monitoring is, as ever, dependent on the surveyor's skill in minimising all possible sources of error, and must not be rushed.

2.4.4 Targets

The angular intersection process requires omnidirectional targets which are easy to bisect. At the Mansion House, because of the short range to the control points, the targets are small ball bearings no larger than pinheads. These are cheap to manufacture, cheap to install, omnidirectional and almost unobtrusive to the naked eye. This final attribute, unobtrusiveness, was the reason why the more expensive intersection technique was adopted at the Mansion House. The targets have to remain for at least 5 years, and as the building is a prestigious tourist attraction and historic monument large conspicuous targets were deemed to be unacceptable.

At the Tower of London the feature actually monitored was the Moat Wall. It was possible to place all the targets on the inside face, which is out of sight of the public. The targets used here were complex and expensive to manufacture and install. As shown in Figure 2.8 they have a retroreflector at

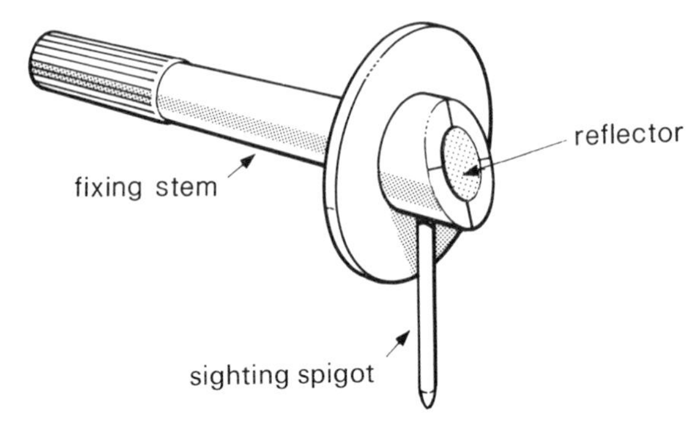

Figure 2.8 Combined target for EDM and theodolite.

the centre and a protruding spigot set to the separation distance between the telescope optics and the EDM optics. The observer therefore sights the end of the spigot and not the centre of the prism. For the total station's EDM to receive sufficient signal back from the retroreflector each reflector has to be pointing back to a control point to within 15° of the line of sight. This was achieved by a combination of careful drilling of holes in the wall and chamfering the head of the target so that it could be rotated in its wall socket to achieve optimum signal return.

At the Mansion House the targets are so unobtrusive that they will probably be left on the structure for all time. At the Tower of London the client insists that on completion of the monitoring scheme the targets must be removed and the socket holes filled so that there is virtually no evidence of the works. To achieve this the design and manufacture costs had to be increased considerably. The sockets in the wall had to be recessed so that they could eventually be covered by mortar. The targets had to be locked into the sockets so that they could not move, even if accidentally knocked, but with a seal that could be broken at the end of the project.

With the intersection technique the size of the omnidirectional targets, usually spheres, can be increased to suit the range to the control and magnification of the observing telescope. Larger ball bearings add little to the costs of the targets and installation is quick because orientation of the target, relative to the control point, is of no significance. Such ball bearing targets, even if rust-free metals such as stainless steel or aluminium are used, cost only a few pence each.

The targets at the Tower of London cost £8 each in 1987. This was relatively cheap because the retroreflectors were one centimetre diameter plastic reflectors, costing 25p each, similar to reflectors used on bicycles. Up to 50 metre range certain total stations, but not all, can derive quality distances from such reflectors. Over 50 metres it would be necessary to use glass retroreflectors. These could cost between £100 and £200 each, are large, and in certain circumstances could be aesthetically unacceptable.

2.4.5 *Choice of monitoring system*

2.4.5.1 Elements of system. From the foregoing examples it is clear that several things have to be taken into consideration in deciding which surveying technique is most appropriate to the structure that has to be monitored. Accuracy, targeting, instrumentation and timing are the key factors, for which the main considerations are given below.

2.4.5.2 Accuracy. As mentioned in section 2.4.3.3, with such sophisticated equipment presently available there is a tendency to offer or specify equipment more advanced than the accuracy requires. The cost–accuracy graph is hyperbolic. Engineers requiring monitoring should assess what degree of structural movement will be truly significant in terms of damage. Measuring small movements which have no effect is a waste of time and money.

Over a long range, distance-measuring solutions tend to produce higher accuracy. The use of modern EDM, bearing and distance is probably the best solution as instrumental error is fairly standard, i.e. it increases by 1–5 parts per million or 1–5 mm per kilometre.

The quality of angular intersection diminishes as the range increases. This is because the observational error for each ray, measured in seconds of arc, is exaggerated with range, and the resultant vectors increase with the tangent of the angle.

If the monitoring points can be occupied then trilateration or triangulation could be appropriate.

Over a short range, all four techniques (section 2.4.1) are viable and, depending on the instrumentation, geometry, etc., can achieve reasonably equitable accuracy.

2.4.5.3 Targeting. The geometrical lay-out of the targets and the available space for the control points should be such that good angular intersections can be achieved. The acceptability of their appearance has to be considered. If targets cannot be fixed to the structure, distance will have to be read by tape or an EDM which does not require a reflector. Equally the intersection principle could be applied, if conspicuous points of hard detail can be viewed from two directions.

In the worst case where the structure to be monitored is inaccessible to affix targets and there are no clearly identifiable points on the masonry then it may be necessary to resort to, for example, shining a laser on a series of random points and then intersecting the laser spot.

2.4.5.4 Instrumentation. The availability of suitable instrumentation is a constraint, as is the budget available for the scheme. The cost of possible premature structural damage or of delay to a construction project must be balanced against monitoring costs.

2.4.5.5 Timing. The frequency with which observations are required to be taken is also an important determinant of an appropriate scheme. If the structure is only required to be monitored at (say) monthly intervals then the highly expensive 'real-time' solution involving electronic instrumentation and computers is probably overkill, assuming that traditional instruments can produce the required accuracy.

2.4.6 Observational errors

2.4.6.1 Relative vs absolute movement. It is fundamental when observing from remote control that the control itself is stable, that all instruments are in good adjustment and that they are set up reliably over the control points.

Often it is impossible to see the monitoring targets from control points outside the sphere of influence of subsidence. This is certainly the case at the Mansion House. On this scheme, if measurement of absolute movements had been required, supplementary control would have been necessary remote from the influence of the Docklands Light Railway tunnelling, but which could be seen from the control points that could see the Mansion House.

It is much more difficult to maintain high absolute accuracy if control has to be brought in, in two stages, and this should be avoided wherever possible. Modern 'real-time' intersection systems, as used at the Mansion House, have another attribute which is not immediately obvious. At the Mansion House surveying was not required to monitor absolute settlement, as other instrumentation had been installed for this purpose. Only relative movement of all parts of the external shell of the building was required to determine whether deformation was uniform or not. If movement was uniform then the structural integrity of the building was likely to remain stable with little likelihood of serious cracking.

The software provided with the electronic system allowed the creation of a three-dimensional spatial model of the 200 monitoring points defining the Mansion House. Subsequent surveys allowed the creation of new models which could be compared point by point or group by group of points. Hence it was possible to assess, almost instantaneously, whether any part of the building was moving in conflict with other adjacent points. If there were no significant differential movements then the building was assumed to be 'floating', intact, as a unit.

2.4.6.2 Calibration. It is common sense that all instrumentation used must be checked and calibrated at the beginning and end of each survey to eliminate instrument errors, or in some cases before and after each set of readings. For the greatest accuracy it is desirable to match individual targets to specific locations.

2.4.6.3 Centring. Centring over control points is one of the greatest sources of observational error. If possible, instruments should be set up at the control points by forced centring, i.e. attached directly to a trivet or bolt on a fixed monument. If this is not possible and tripods have to be used, centring may have to be established using a monitoring quality optical plummet and not the standard optical plummet found in most tribrachs.

2.4.6.4 Thermal effects. The sun can have considerable effects on observations:

(1) Using conventional theodolites with liquid bubbles, levelling of the instruments can be affected.
(2) Tripods should not be set up on tarmac as they can become unstable in hot conditions.
(3) The tripods themselves can expand causing dislevelment errors.

2.4.6.5 Overall considerations. In monitoring, every aspect of the means of measurement must be given in-depth consideration to minimise observational discrepancies that could diminish the overall quality of the results. The Mansion House is a very difficult location in which to attempt to achieve submillimetre accuracy. Vibrations from road traffic, Docklands Light Railway tunnelling and underground trains can be detected. Further to this, considerable shimmer is experienced from the heat of the sun and vehicle exhausts. Pedestrians and parked or moving vehicles are a minute by minute hazard.

2.4.7 Data recording and tabulation of results

In this computer age data can be processed in 'real-time' on-site or by computer in the office. Obviously the 'real-time' instrumentation automatically records the data. These data can be transmitted immediately on completion of observations to any party wishing to evaluate the data.

With all non-automatic systems, it is sensible to record the data in electronic data loggers which permit direct transfer to office computers, thus eliminating keyboard inputting time and human errors.

Computers can be used not only for calculation of coordinates but also for tabulation of results. Using spreadsheets one can automatically derive tables of results with comparison of recent movements and overall movements, as in Table 2.2. Equally, computers can be used to produce vector movement graphs or settlement contour diagrams.

Most monitoring schemes require that results are evaluated as soon as possible so that the cause of movement can be stopped at short notice. In this context computers are invaluable compared with the time needed for manual computations, tabulations and comparison. By presenting only essen-

Table 2.2 Example of spreadsheet to show recent and overall movements. Docklands Light Railway—surface and building monitoring task: Tower of London, Moat Wall—three-dimensional coordinates

Survey no. 49 Survey date: 20/11/89

Point ID	Survey no. 26			Survey no. 48			Survey no. 49			Overall difference			Latest difference		
	Eastings	Northings	Level	Eastings	Northings	Level	Eastings	Northings	Level	Difference east	Difference north	Difference level	Difference east	Difference north	Difference level
AOU	556.207	633.972	107.222	556.206	633.975	107.222	556.209	633.974	107.218	2	2	−4	3	−1	−4
AOM	556.214	633.711	105.574	556.213	633.712	105.573	556.217	633.712	105.571	3	1	−3	4	0	−2
AOL	556.231	633.408	103.992	556.231	633.408	103.994	556.235	633.407	103.992	4	−1	0	4	−1	−2
A1U	558.644	634.344	107.194	558.644	634.344	107.195	558.647	634.344	107.192	3	0	−1	3	0	−2
A1M	558.630	634.303	105.554	558.629	634.303	105.554	558.633	634.300	105.552	3	−3	−2	4	−3	−2
A1L	558.652	634.168	103.955	558.650	634.170	103.956	558.655	634.166	103.954	3	−2	0	5	−4	−2
A2U	560.665	634.552	107.220			107.217			107.217			−3			0
A2M			105.534			105.534			105.533			−2			−1
A2L	560.673	634.393	103.943			103.945			103.943			0			−2
A3U	562.633	634.761	107.234	562.632	634.762	107.233	562.636	634.762	107.232	3	1	−2	4	0	−1
A3M	562.631	634.749	105.596	562.630	634.748	105.596	562.634	634.747	105.594	3	−2	−2	4	−1	−2
A3L	562.644	634.622	103.993	562.643	634.621	103.994	562.647	634.618	103.993	3	−4	0	4	−3	−2
A4U	564.612	635.003	107.280	564.615	635.000	107.280	564.618	635.001	107.278	6	−2	−2	3	1	−2
A4M	564.606	634.959	105.640	564.607	634.961	105.639	564.611	634.960		5	1		4	−1	
A4L	564.618	634.855	103.992	564.617	634.855		564.619	634.854	103.991	1	−1	−1	2	−1	
A5U	566.608	635.243	107.218	566.609	635.244	107.218	566.610	635.245	107.217	2	2	−2	1	1	−1
A5M	566.588	635.193	105.566	566.588	635.193	105.565	566.589	635.193	105.564	1	0	−3	1	0	−2

tial results in a simple format the task and reliability of interpretation are facilitated and enhanced.

2.5 Movement across cracks

2.5.1 Present condition survey

It is not appropriate here to discuss in any detail the measurement of the size and extent of cracks for a survey of present condition, but an assessment of the general nature and disposition of cracks as a whole in a building, or at least the part of it to be studied, is an essential prerequisite to deciding which should be monitored in detail. The value of photogrammetry for record purposes should be considered at this stage (see chapter 3). It may also provide the means of measuring crack size and movement if accessibility to the surface of the building is limited or difficult.

To the extent that a clear overview of the size, extent and disposition of cracks enables a good assessment to be made of the behaviour of the building and the likely underlying causes of damage, the selection of cracks to be monitored in more detail to follow continuing movement will be productive of reliable data focused on the real issues.

Comprehensive guidance on 'present condition' surveys is given in BRE (1989b) where accuracies of 1 mm are generally considered sufficient. Recommendations are made based on the use of graduated rulers, which should be of matt, chrome-faced steel graduated in full millimetres on both edges of one side. Techniques are suggested for recording both the numerical measurements and also observations about the nature of each crack recorded e.g. tapers, steps, shear and compression.

Weathered edges and dirt may be signs of a crack of some age, possibly no longer active. Other signs of movement, such as pipes, parapets and distortion of doors and windows may be diagnostic and should be recorded. Planned systematic procedures for recording information will ensure its later correct interpretation, and transcription should be avoided as far as possible to prevent the introduction of errors.

The significance of such measurements and observations in terms of the urgency and nature of any repair may be assessed against criteria given in BRE (1975; 1985).

2.5.2 Measuring movement

Measurements of movement across cracks will often need to extend over months, or even years, before a valid assessment of performance can be made. Therefore the need for reliability, ease of use and robustness discussed for levelling stations are applicable. They are reflected in the methods to be preferred in order to achieve the accuracy of 0.1 mm which is appropriate for this type of measurement.

As indicated in the previous section, selection of cracks for further monitoring should be based on the results of a present condition survey of sufficient extent. Generally the largest active cracks should be selected because they may show the greatest rate of movement and therefore yield more information in a shorter time than otherwise.

Judgement and experience will always play their part, but other criteria suggested by BRE (1989b) for selecting cracks are:

(1) The whole of the cracked area of the building should be represented in the cracks monitored.
(2) They should be chosen so as to reveal distortion which may occur and may be characteristic of causes of movement, e.g. hogging.

The accuracy of measurement required to ensure that real movements are being monitored is 0.1 mm. This accuracy effectively rules out the use of tell-tales, be they plain glass or more sophisticated graticules, quite apart from their vulnerability. Tell-tales may, however, still be useful as advance warning devices particularly, for example, if watched by an occupier who is briefed to alert a professional under appropriate circumstances.

A readily available device of long-standing use is the demountable mechanical strain gauge, or Demec gauge. Its resolution is 0.002 mm with a total range of 4 mm so it is only relevant to the study of small movements in great detail and care should be taken in the interpretation of what might seem to be rapid or large movements. The reference points are cheap, unobtrusive, small metal discs, with a drilled location hole, stuck to the building.

A preferred method for monitoring cracks in a plane surface is to take measurements with a caliper gauge between small screws set in the wall or other building element. Although they need to protrude about 5 mm they are relatively unobtrusive, rugged and usually free from disturbance. A digital caliper is desirable as it is less likely to be misread, but it should be capable of sitting against the building surface so that the jaws bear on the shanks of the screws. An electronic device also has the potential for being used with a portable digital recorder and computer processing of the data.

Two screws are sufficient to measure movement across a crack if the screws can be set with certainty in the line in which movement is occurring or expected to occur. Otherwise three screws are required so that magnitude and direction of movement can be measured. Detailed procedures for setting out the positions of the screws and installing them, and for taking measurements and recording them are given by BRE (1989b). The techniques can be adapted using the same type of calipers to monitor cracks at internal corners and steps at cracks. A detailed photographic record of the cracked areas can be a valuable aid to interpretation away from the building. If the use of screws is unacceptable various optical techniques using a magnifier with graticule may be applied with small hand-made pencil marks.

References

BRE (1975) Failure patterns and implications. *Digest 176*. Building Research Establishment, Watford.

BRE (1985) Assessment of damage in low-rise buildings. *Digest 251*. Building Research Establishment, Watford.

BRE (1989*a*) Simple measuring and monitoring of movement in low-rise buildings. Part 2: Settlement, heave and out-of-plumb. *Digest 344*. Building Research Establishment, Watford.

BRE (1989*b*) Simple measuring and monitoring of movement in low-rise buildings. Part 1: Cracks. *Digest 343*. Building Research Establishment, Watford.

Burland, J.B. and Hancock, R.J.R. (1977) Underground Car Park at the House of Commons; Geotechnical aspects. London. *The Structural Engineer*, **55**(2), 87–100.

Burland, J.B. and Moore, J.F.A. (1973) The measurement of ground displacement around deep excavations. *Proceedings of the BGS Symposium on Field Instrumentation*, Institution of Civil Engineers, London.

Cheney, J.E. (1973) Techniques and equipment using the surveyor's level for accurate measurement of building movement. *Proceedings of the BGS Symposium on Field Instrumentation*, Institution of Civil Engineers, London, and Building Research Establishment Current Paper CP 26/73, Building Research Establishment, Watford.

Cheney, J.E. (1980) Some requirements and developments in surveying instrumentation for civil engineering monitoring. *FIG Commission* 6 *Symposium on Engineering Surveying*. University College, London.

3 Photogrammetry

D. STEVENS

3.1 Introduction

Most people readily associate photogrammetry with its use in the production of maps and plans by aerial surveying techniques. Whilst this remains the main use for photogrammetry the overall range of applications extends from orthodontic measurement to the three-dimensional modelling of nuclear reprocessing facilities.

For its use in the measurement of buildings and structures it is necessary to go back to the mid-nineteenth century when the Prussian, Albrecht Meydenbauer (1834–1921) recognised the possibilities of producing elevational drawings from photographs. In 1867, some 32 years after Fox Talbot had developed his paper negative, Meydenbauer produced some remarkable drawings of the Church of St Mary Freyburg-a.-Unstrut (see Figure 3.1) and, coincidentally, coined the term photogrammetry.

Figure 3.1 St Mary's Church at Freyburg-a.-Unstrut. This elevation of the north facade was produced by Albrecht Meydenbauer in 1867 using graphical photogrammetric techniques. (Reproduced with the permission of Carl Zeiss, Germany.)

The precise art of photogrammetry has continued to serve building engineers and architects, with its noted development by Professor Hans Foramitti (1923–82), in Austria, and Professor E.H. Thompson (1910–76) in the UK. Thompson's work at Castle Howard in 1962 is of particular interest as an analytical approach was taken to resolve the problems of using photography not taken for measurement purposes.

In recent years the use of photogrammetry in architecture has been furthered by the Institute of Advanced Architectural Studies under the direction of their Chief Surveyor, R.W.A. Dallas, who has been personally responsible for overseeing the recording, by photogrammetry, of many important buildings under the control of English Heritage.

In the commercial sector, much use is made of photogrammetry by architects and developers during the initial stages of refurbishment projects. Not only does photogrammetry provide the essential accurate elevational drawings but it can also provide accurate horizontal and vertical profile data. These drawings, whether produced as conventional transparencies or as computer-aided design (CAD) files, provide the architect with the necessary data for the redesign, as well as providing a form of insurance should any damage occur during the reconstruction work. This form of insurance is equally appropriate to facade retention projects. Should a facade collapse, the evidence in the form of a photogrammetric survey will permit accurate reconstruction to take place.

Ecclesiastical architects and building archaeologists also make extensive use of photogrammetry to assist in the reconstruction or maintenance of our building heritage. Many facades of historic and important buildings have been recorded, including the cathedrals at Lincoln, Lichfield, Salisbury and Wells.

September 1989 saw a tragic fire which gutted Uppark House in Sussex. By Christmas that year, the exterior and the majority of the interior had been recorded by photogrammetry as the first stage of the rebuilding programme. The results of the recording work at Uppark are being used by archaeologists to analyse the history of the construction of the building, as well as by the architects and structural engineers whose task it is to determine the precise method of reconstruction.

Figures 3.2, 3.3, 3.4 and 3.5 show typical examples of both the internal and external recording taken from the Uppark project. The use of photogrammetry at Uppark is of additional interest because of the intensive use of computers to store the recorded data in three dimensions and, where appropriate, to use this three-dimensional data to assist in the accurate reconstruction. In particular, vertical profiles of severely damaged parts of the south facade have been produced, by photogrammetry, to measure the extent of the distortion in the wall caused by the fire.

Photogrammetry, as a matter of course, records all data as a series of three-dimensional coordinates, and whilst the third dimension is often ignored

Figure 3.2 Uppark House, Harting, Sussex, England. This is one of the photographs taken of the east face using a Zeiss UMK 10/1318. (Reproduced with the permission of the National Trust; Architects, the Conservation Practice.)

Figure 3.3 Uppark House, east elevation. This drawing was produced by AMC Ltd using a Galileo Siscam Digicart 40 photogrammetric instrument. All final plots on the Uppark project were computer generated using AutoCAD. (Reproduced with the permission of the National Trust; Architects, The Conservation Practice.)

Figure 3.4 Uppark House, salon. One of the UMK photographs used in the preparation of the elevation shown in Figure 3.5. The ground floor salon was mostly undamaged, unlike the first- and second-floor levels. (Reproduced with the permission of the National Trust; Architects, The Conservation Practice.)

because of problems in portraying it in graphical form, it makes a natural partner for today's three-dimensional CAD systems.

The Uppark examples demonstrate the level of data that can be recorded entirely by photogrammetric procedures.

3.2 Principles

In order to appreciate the application of photogrammetry in monitoring, it is useful to gain a rudimentary insight into the basic technique.

A simple definition of photogrammetry is: 'The precise art of abstracting measurements from imagery (photography)'.

Most people will have an impression of photogrammetry in terms of pairs

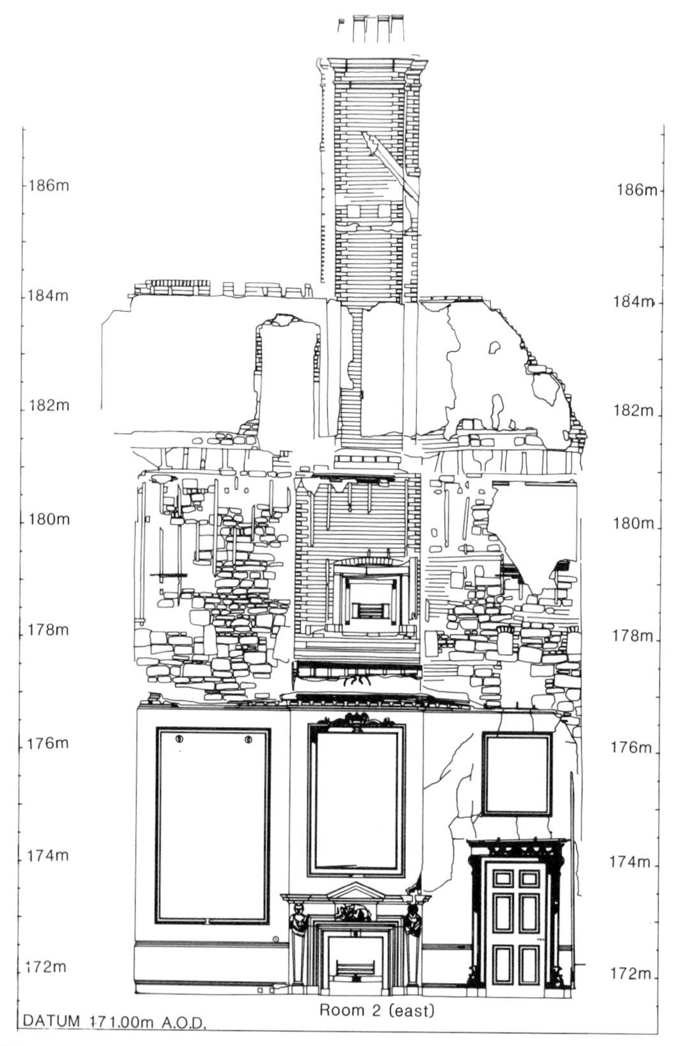

Figure 3.5 Uppark House, salon. This elevation shows the level of detail required by archaeologists to help determine the history of the construction. (Reproduced with the permission of the National Trust; Architects, The Conservation Practice.)

of photographs and complicated stereoscopic instrumentation. Certainly in most cases a pair of photographs is taken, with a separation between the camera positions. Conventionally, the cameras will be aligned so that the principal axes are parallel (see Figure 3.6). This allows the photographs to be viewed through a stereoscope and the 'stereo' or three-dimensional image to be observed.

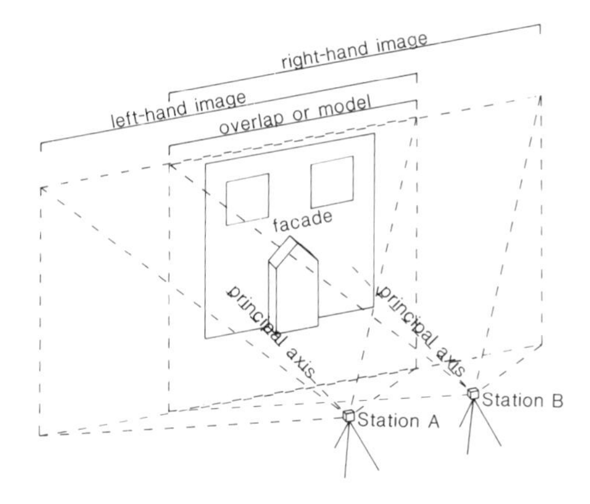

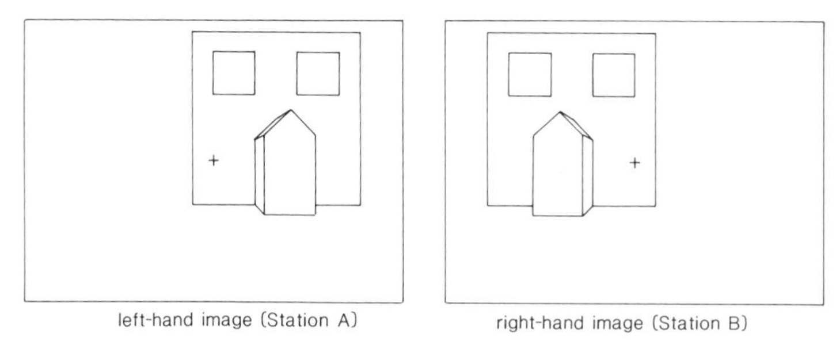

Figure 3.6 Conventional parallel photography. The camera stations are positioned to create acceptable geometry in the most economic manner. The principal points (optical centres) are indicated by the cross on the two images.

The photography is the most important part of the photogrammetric process. Later on it will become evident that if the physical relationship between the camera positions and the object is optimised, then the best, i.e. most accurate, results will be obtained. Poor geometry will result in a poor solution.

To photogrammetrists a photograph is obviously much more than a simple snap shot. The camera can be equated to the basic surveying instrument, (namely the theodolite), and the photograph can thus be equated to the series of angles observed from a single station.

By measuring the base distance between two theodolite stations the surveyor is able to relate and orientate his station positions and produce a plan to the required scale. One criterion for the surveyor is to ensure that his

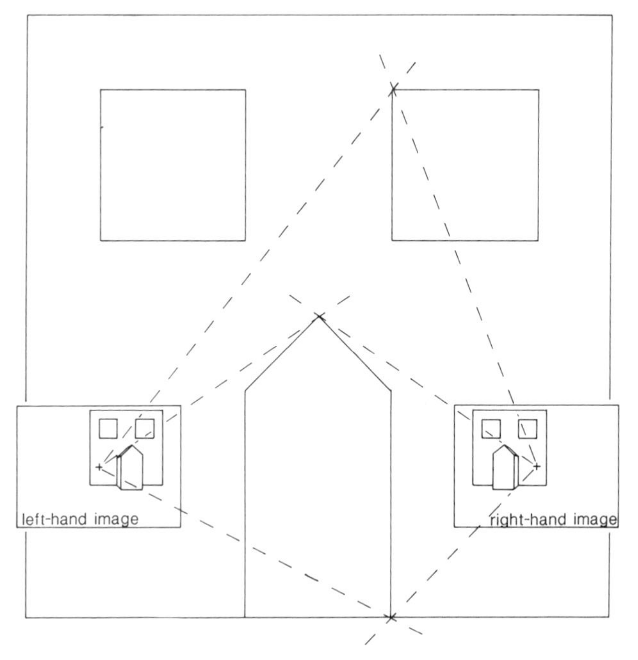

Figure 3.7 Intersection geometry. By aligning the two images through the principal points and their homologues, an elevation can be graphically constructed. Extended rays from each principal point, through the point of detail, create an intersection representing the relative position of that point of detail to the selected scaled distance between the two images.

theodolite is properly levelled (horizontal). If the photographer is able to set his camera so that it is normal to the object, i.e. the principal axis is normal to the facade to be measured, then it is possible to produce an elevational drawing using simple intersection geometry (see Figure 3.7). In 1867 this was probably the only method that Meydenbauer had available to him.

This simple intersection geometry produces two-dimensional data, there being no depth information available. However, by looking again at the left- and right-hand images (Figure 3.8) it will be seen that, whilst the main part of the recorded facade appears to be flat, the depth of the porch is recorded because of the perspective properties of each photograph. Whilst point B appears in front of point A on the elevational drawing (Figure 3.7), it appears displaced on each image. The intersection geometry removes the effects of this displacement (parallax) as it is radial to the principal point (optical centre) of each image. However, this parallax provides the information needed to determine the third dimension, as the difference between $A–A_1$ and $B–B_1$ is in proportion to the actual distance AB. Thus the third dimension is derived from the effective difference in the x-axis.

Deriving x and y data by simple intersection geometry and the z data by

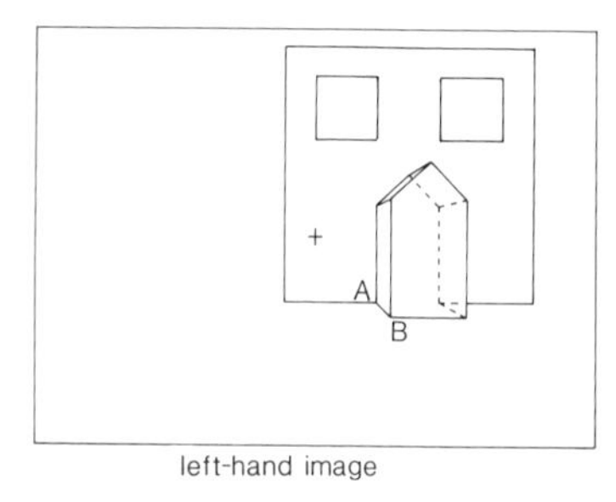

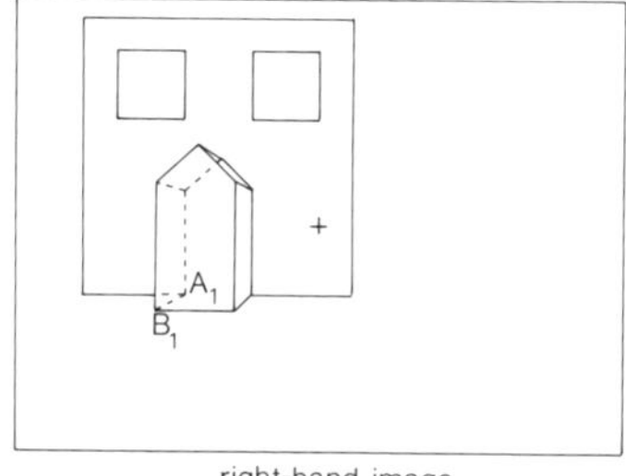

Figure 3.8 x-Axis parallax. The effect of depth in the image creates parallax or displacement which is radial to the optical centre of the image. Although points A and B have the same x and y coordinate values, they differ in z, and this is represented by the difference, $(A_1 - A) - (B_1 - B)$.

parallax measurement is the basis for many of the 'analogue stereo plotting instruments' designed and built up to the end of the 1970s. The analogue instrument, such as the ubiquitous Wild A8, helped to map many thousands of square kilometres of the earth's surface, and these instruments will continue to be used while the demand for conventional mapping exists. However, the analogue instrument must be regarded as yesterday's technology and is no longer adequate for the three-dimensional measurement tasks, or standards of precision, which photogrammetry is now expected to achieve, especially with non-topographic subjects.

The analogue instrument had many limitations which, whilst of no major importance in the production of maps and plans, were sufficient to prevent its use in a fully three-dimensional environment. The more important of these limitations were as follows:

(1) Restrictions in camera rotations. The principal axes of the camera had to be within a very few degrees of normal to the plane of the object, otherwise they would fall outside the mechanical restraints of the instrument.
(2) Restrictions on the range of focal length. Again, owing to the mechanical linkages only a small range of focal lengths could be accommodated.
(3) No means of using imagery from non-metric cameras (see section on cameras, 3.4.5).
(4) No instrument management or analysis of orientation data.

The term analytical photogrammetry had been penned earlier when progressive photogrammetrists had realised that it was more precise to compute the coordinate data mathematically rather than to use mechanical devices to derive and graphically plot the data. The relative reduction in the cost of computers during the late 1970s and early 1980s prompted the development of the analytical instrument as it is now known. It must be appreciated by the reader that any publication on a technique such as photogrammetry is forced out of date by continuing development. However, it is important to relate to the current levels of technology, and we must be grateful that the relative cost of computing has permitted the development of the analytical instrument and expanded the use of photogrammetry through much greater flexibility and increased precision.

To obtain results by computational photogrammetry, two stages are involved following the gathering of the photography and control on site. The first of these stages is to calculate the positions and orientation of the cameras relative to the subject. The second stage involves using this information to derive the three-dimensional coordinates of the point or points of interest.

The first stage is known as the 'orientation', to which there are conventionally three steps.

Step one can be regarded as the inner orientation, which relates the photograph geometry to the measuring system of the photogrammetric instrument. This involves recording the position of the fiducial points which are recorded on the photographs (see Figure 3.2). These fiducial points are exposed onto the film either by direct exposure or by projection depending on the camera design. They are positioned to allow the principal point (optical centre of the lens system) to be derived, and are to determine any dimensional changes that may have occurred in the film between the acquisition of the photography and the photogrammetric measurement.

Step two is to determine the three-dimensional relationship between the two photographs. Within the common overlap between the images (sometimes referred to as the model as it is the area that can be viewed in three dimensions) a minimum of six points which can be accurately read on both of the photographs are identified. Although the computer does not know the actual positions of these selected points they are sufficient for it to compute the three-dimensional relationship between the two photographs and thus derive the relative camera positions. This step is known as the relative orientation.

Step three of the orientation exercise is the absolute orientation, which adds scale and a datum into the equation. Under optimum conditions the photography will include a number of points with known values which permit the software to compute the camera positions on the chosen coordinate system. However, if conditions permit, this 'control' can be as simple as the inclusion of a scale bar (such as a levelling staff) within the photographic overlap.

Most analytical software packages also contain the alternative option of carrying out all the orientation steps as a single computation. This involves the technique known as 'bundle adjustment'. The term 'bundle' refers to the bundle of light rays passing from the object, through the lens system of the camera to the focal plane.

The selection of which orientation procedure to adopt is governed by the geometric relationship between the object, the camera stations and, in particular, the control.

Upon completion of the orientation procedures, the computer has sufficient data to permit measurement of the points of interest. Conventionally the operator will view the 'stereoscopic model' (or three-dimensional image) by mentally fusing the two images into one. The points are measured by placing the 'floating mark' on the point of interest and recording the coordinates by touching a foot switch or similar. The 'floating mark' is in reality two marks, one being positioned in each of the two optical viewing chains. In the same way that the two photographs appear as a three-dimensional model, the two marks are mentally fused and appear to float above or below the surface of the object. The skill of the photogrammetrist is to position the fused 'floating mark' so that it appears to be coincidental with the point to be measured. This is done by effectively increasing (to lower) or decreasing (to raise) the distance between the two measuring marks. The instrument is actually measuring the amount of x (major) and y (minor) parallax which affects the apparent position of the floating mark.

From the data thus gathered, the computer software is able to calculate the position in space of the point being measured from the x and y photo coordinate values of the two measuring marks. Output can be to a graphics screen, a plotter, tabulated to a line printer or stored on magnetic media.

Analytical techniques are best suited to non-standard applications of photogrammetry (which includes most forms of monitoring) so it is important to look at the additional flexibility that this form of photogrammetry offers.

Previously mentioned are the restrictions imposed by the analogue solution in terms of camera rotations. For conventional aerial surveys and even most architectural (elevational) surveys, parallelism of the principal axes (see Figure 3.9) produces adequate geometry and a good stereoscopic image which is easy to view and produces acceptable results. However, as soon as an increase in accuracy is required, the geometry needs to expand and it is no longer possible to guarantee parallel principal axes. With the analytical approach, convergence of the principal axes can be introduced, thereby increasing the camera base to object distance ratios and thus improving the precision of the operation (see Figure 3.10). It will be noted that, by introducing convergence, the camera stations can also move closer to the object, adding further precision by decreasing the scale of the photography.

In quantitive terms, if the building in Figures 3.9 and 3.10 is, for example, 20 metres wide and of similar height, the convergent solution would produce results approximately two and a half times more precise than the results of the parallel solution.

3.3 Applications

As this book demonstrates, there are a variety of techniques for use in the monitoring of buildings and structures. Photogrammetry is one of those

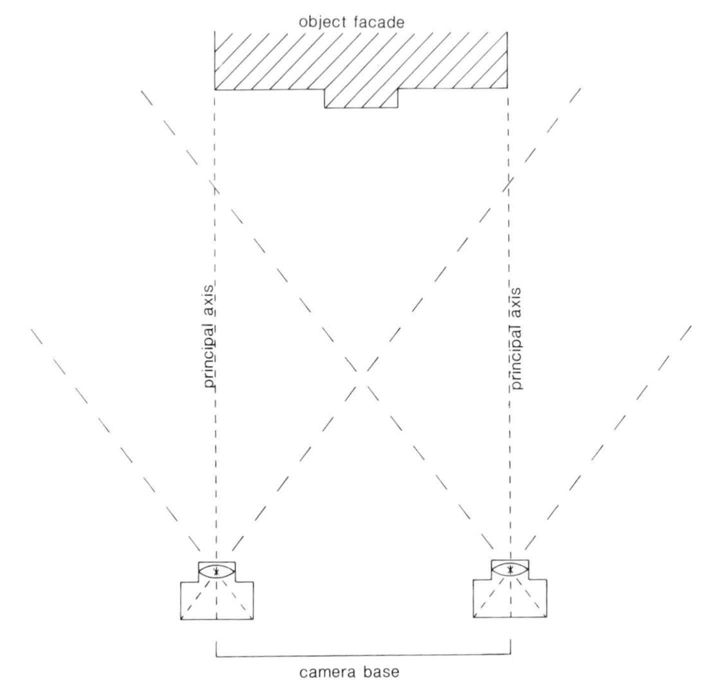

Figure 3.9 Camera geometry—parallel option. The cameras are sited to create the optimum achievable ratio between the object distance and the camera base. In this case a ratio of less than 2 : 1 has been achieved through the use of a wide-angle lens system.

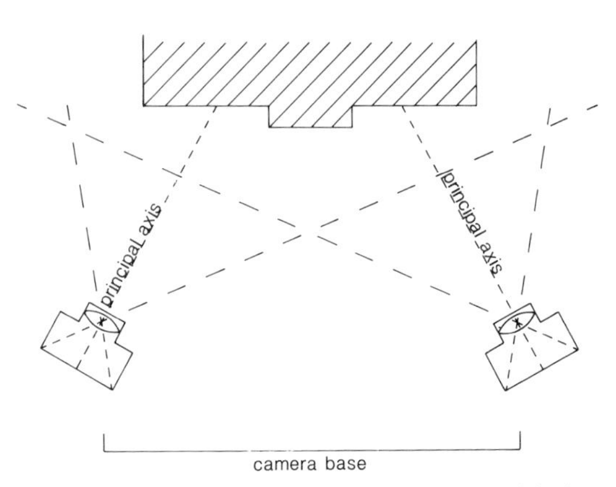

Figure 3.10 Camera geometry—convergent option. Compared with the parallel option, the object distance to camera base ratio has reduced from approximately 2 : 1 to nearer 1 : 2, permitting a 90° intersection angle over the centre of the facade, diminishing towards the edges.

techniques, and whilst it can be used as a direct alternative for some of the other options, some situations arise which take full advantage of its unique attributes, sometimes making photogrammetry the only viable option. Below are discussed the more beneficial applications of photogrammetry:

(1) The use of photogrammetry on a project to produce elevational drawings logically suggests that photogrammetry should also be used for any monitoring that may be required. This is because the equipment and specialist personnel are already available.
(2) It will be shown later that it is often preferable to target points to aid identification and accuracy. However, situations arise where access to the building concerned is not feasible, possibly because of physical constraints, such as water, or hazardous constraints such as radioactivity or heat. In these situations photogrammetry can often produce acceptable levels of precision. In theory, photogrammetry is classed as a remote sensing technique and much use is made of this particular attribute.
(3) The previous mention of water was given as an example of inaccessibility, but situations occur when the monitoring has to be carried out from a water-borne platform, such as a barge. Other unstable or dynamic platforms include hydraulic access platforms (cherry-pickers) and even scaffolding towers, if not fully braced. In these conditions the effect of the frozen action created by the fast shutter speed of the camera can combat the disadvantages of a moving or unstable platform and thus permit the monitoring to be undertaken.
(4) There are occasions when the object building moves, such as will be described later in the case history of Merritt Point. In these situations the frozen action effect of the photography permits the recording of a series of scenarios during the period of measurement.
(5) A photograph records an infinite amount of data and thus photogrammetry is the most appropriate tool for the monitoring of buildings where a high density of information is required.
(6) In certain types of monitoring projects it is not always possible to forecast with any accuracy exactly how a facade will respond when put under stress, whether by increased physical loading, by partial demolition or by undermining, etc. In these circumstances it is not possible to direct the monitoring to the specific area where movement actually occurs.

By using photogrammetry, not only can areas of movement be examined using a high density of measured points, but retrospective measurement can also be carried out using the photography from the previous photographic recording.

3.4 Techniques

3.4.1 Planning

Once an understanding of the principles of photogrammetry and the most beneficial areas of usage has been reached, it is a logical progression to consider the actual application of the techniques. Although one project may be similar to a previous one in some ways, each is unique and it is only possible to give guidelines as to the optimum way in which a project should be approached.

The planning stage of any project cannot be skimped by making assumptions which inevitably prove to be wrong. Later on the planning stage will be discussed in detail, but there is one point to be made that is sometimes forgotten. In considering monitoring projects, the first criterion has nothing to do with the measuring technique, but everything to do with the performance of the object building being monitored, and thus the purpose for which the monitoring is required. The standards of precision requested must be considered not only against the practical limitations of the project but also against levels of movement expected in the structure.

3.4.2 Site recording

The purpose of the photography is to record the object, or that part of the object, to be monitored or measured. Each part of the object must be recorded from at least two camera stations with appropriate base distances between them. The photography must also incorporate any control positions and must meet any other criteria set down at the planning stage.

Having carried out a detailed analysis of the requirement, the photographer will know where to position the camera stations for optimum accuracy and coverage. However, before proceeding with the photographic recording there are two preliminary stages of work to be undertaken.

3.4.3 Targeting

In monitoring projects it is most likely that convergent photography will be employed to obtain the required accuracies. With more than 10–15 degrees of convergence between the principal axes it becomes difficult for the photogrammetrist to fuse the stereoscopic image and he therefore has to rely on positioning each of the two component measuring marks independently. Added to the problem of fusing the images, the obliqueness of the photography alters the appearance of the detail points, making it difficult to identify the same points precisely in the two images.

Consequently it becomes important to target the points to be monitored if at all possible. Different types of targets are readily available and target selection needs careful consideration. The minimum and maximum scales of

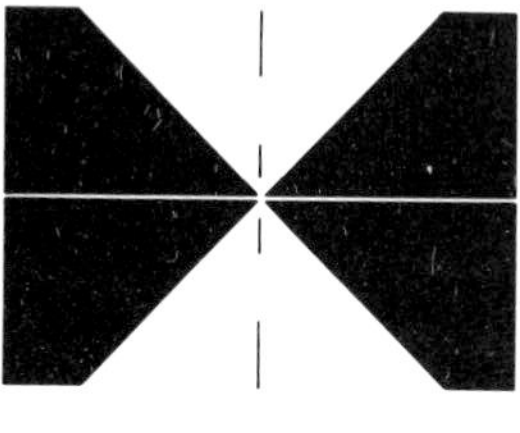

Figure 3.11 Target types.

the photography must be taken into account, as well as the relative size of the measuring marks in the photogrammetric instrument.

Figure 3.11 portrays the two most common designs of target, the quadrant and the bull's-eye, both of which are designed to be used over a range of distances. Both designs are available on flexible plastic material, which can be applied using an epoxy resin for permanent siting. The bull's-eye targets are also available on a waterproof paper which is ideal for temporary siting, with recommended use of a polyvinyl alcohol (PVA)-type adhesive on porous surfaces. Whenever targets are applied they should be numbered clearly to avoid any confusion.

Another option for targets is to use retroreflective material such as is produced by the 3M Corporation. Targets printed onto this Scotchlite material are recorded by flash illumination, which whilst too weak to illuminate the building permit the reflected light to be recorded on the photographic emulsion. One unusual advantage of using these targets is that they permit multiple exposures, thereby illustrating dynamic movement. However, this is not always a practical alternative because of ambient light levels overexposing the film.

The positioning of the targets, and the numbers involved, is normally stipulated by the structural engineer. With buildings it is sometimes possible to position the targets through window openings, but often a 'cherry-picker' is needed. When positioning the targets, their attractiveness to the local vandals must also be considered, and they should be positioned well out of reach.

There will be occasions when targeting is not possible, or is not essential. With some buildings there are sufficient hard detail points to permit the use of convergent photography, whilst in other cases the shape and size of the building, or the accuracy requirement, allow for conventional 'parallel' photography to be used.

3.4.4 Control

Progressive photogrammetrists often argue the case for minimising the control requirements by the simple addition of adding scale to the bundle adjustment solution. In effect, instead of first orientating the model in the normal way, and then abstracting the point coordinate data, the points to be monitored are included in the orientation procedure. This is in many ways a very tidy solution and can be used to combine the information contained on a number of photographic images, not just to cover a greater area but also to increase the accuracy. However, many monitoring projects relating to buildings and structures are not independently contained and their relationship in absolute terms is required. This introduces the need to add the coordinate values of known points into the absolute orientation stage.

The conventional approach to photogrammetric control is to position additional targets within the area of the photographic overlap, but as far as possible outside the area for monitoring or measurement. The three-dimensional values of these additional targets are usually measured using theodolite intersection techniques, and the reduced observations rotated to a plane parallel to the facade being measured.

Whilst three points in a model will permit the orientation procedures to be completed, it is preferable to include a degree of redundancy in order to permit a check on the accuracy of the orientation. It is also important to ensure that, whenever possible, the control target positions fully encompass the area for monitoring in all three axes.

The Merritt Point example described later in this chapter demonstrates how the control was positioned as far outside the critical area as practically possible, but there are many constraints which need to be considered, again making the planning stage so important.

In the days of analogue photogrammetry, the accuracy of the control was only questioned if the model did not set properly in the instrument. With analytical photogrammetry the orientation procedures produce quantitative data on the accuracy of the match to the control. This gives the photogrammetrist the option of weighting the control values, not only on the indicated accuracies of the control observations, but also on the position of the control point relative to the camera stations. Thus, the control network adjustment needs to include indicative data on the accuracy of each point recorded.

The coordination of the control targets is usually carried out by theodolite intersection, preferably from two linked electronic 1-second or subsecond instruments. The baseline between the instruments will normally be measured by steel band or calibrated electronic distance measurement (EDM) equipment. The baseline, being greater than the width of the structure being monitored, can normally be measured to within a satisfactory level of accuracy by these methods. If more than one facade of a building or structure is involved then the baseline may become part of a traverse, if this is appropriate.

The indicative accuracy value required for each control point can be obtained by computing the height of each point from both stations and comparing the difference (dZ value).

When the surrounding land is unstable, then normal precision surveying techniques will need to be used to ensure repeatable accuracies on future occasions. This will involve traversing the control into the immediate area from the nearest stable land-based survey station or bench mark.

Conventionally the control will be orientated parallel to the longest facade involved. This is done as a matter of convenience to both the photogrammetrist and the engineer.

3.4.5 Photography

There are three types of camera used in photogrammetry and these are known as metric, semi-metric and non-metric. A metric camera is one designed for measurement purposes, whilst a semi-metric camera is one designed for normal photographic usage but adapted, usually by the manufacturer, to make it usable for measurement purposes. A non-metric camera is a conventional camera with minor or no modifications and is generally not suitable for use in precision photogrammetry.

The criteria for photogrammetric cameras are as follows:

(1) Rigidity of the body to permit accurate calibration, and retention of calibration of the lens system relative to the focal plane.
(2) Ability to ensure flatness of the photographic emulsion, as any unflatness introduces serious errors throughout the measurement cycle, and repeatability of the position of the emulsion surface in subsequent exposures.
(3) Facility to project critical points (fiducial marks) onto the photographic emulsions.

Over and above these criteria is the preference to have a lens system which provides high resolution, in terms of recordable lines per millimetre, but minimum distortion (aberrations). It is also necessary to have a lens system with a comparatively wide angle of view, as this allows for the optimum geometry to be achieved with a minimum of convergence in the principal axes.

For monitoring purposes it is also usually necessary to use a large-format metric camera such as the Wild P31 or the Zeiss 1318 UMK. Both of these cameras have standard focal lengths in the region of 100 mm, the Wild having a $5'' \times 4''$ film format and the Zeiss $5'' \times 7''$. The smaller format of the Wild camera is partially compensated by the offset lens, which permits optimum utilisation of the format area in most situations. Figure 3.12 shows the Zeiss camera, but both cameras have advantages to their credit, and both are admirably suited to monitoring work.

In dealing with cameras with relatively large film formats it is not possible to obtain a wide range of emulsions. The P31, with its optically flat reseau plate permits the use of film, and together with its more common format size

Figure 3.12 Zeiss UMK 10/1318 metric camera, the standard Zeiss camera, which has an angle of view equivalent to a 20-mm focal length lens on a '35 mm' camera. The black box on the side is a non-standard power supply used to illuminate the fiducial marks visible on the rear view.

is open to a wider range of emulsions than the UMK. Where possible the lowest practical emulsion speed should be selected to give the finest grain result.

Fine-grain developers such as Ilford Perceptol can help in halving the speed of the emulsion. However, some projects demand that the photography is taken from a boat, cherry-picker or scaffolding platform, and in these situations it becomes essential to select a faster emulsion speed in order to combat camera movement during exposure. In these cases film such as Ilford HP5 is ideal and can be uprated using the appropriate chemistry.

The Zeiss camera, without a reseau plate, is often used with glass plates, and here the range is very limited because of the relatively low demand. The most common emulsion is the Agfa Aviphot Pan 100, which gives very acceptable results under most normal conditions.

As with all forms of photogrammetry, if the quality of the photography is poor, then the final results will suffer. Even with a fully targeted monitoring project, poor-quality photography will create difficulties for the photogrammetrist in trying to position the measuring mark. The quality of the photography goes well beyond the actual recording of the imagery to the good husbandry of the dark room. The negatives, whether glass or film, must receive minimum handling, especially to the emulsion, and must be subject to developing, fixing and washing at constant temperatures, not only to ensure correct processing, but also to avoid any reticulation (minute cracking of the emulsion caused by sudden expansion or contraction of the film base).

All photography should be archivally processed using a hardener in the fix and substantial washing to avoid subsequent discolouration. The film should be allowed to dry naturally after a final wash in a wetting agent to avoid any unsightly drying marks.

3.4.6 Photogrammetric data extraction

Most other forms of monitoring and measurement are, in general, undertaken on site with minimal in-house work to finalise the data. With photogrammetry the opposite is the case. With normal 'plotting' projects, where the end result is an elevational drawing, one day on site may result in 2 weeks' work in the office carrying out the photogrammetric data extraction. The ratio is generally much less with monitoring work as the end result is normally a tabulated list of three-dimensional coordinates, although it is usual to include a plot showing the basic building outline with the target positions marked thereon.

By this stage of the exercise, the control surveyor will have computed the coordinate values and the photographer will have processed the negatives and probably produced some contact prints. The photogrammetrist will need to make himself familiar with the project, as it is unlikely that he will have visited the site or building in question. He will discuss with his colleagues his approach, taking into consideration the positions of the camera stations and the quality and quantity of the control data.

The control, having been automatically recorded and computed without any manual re-entry, will, preferably, be loaded directly into the computer connected to the photogrammetric instrument to form the control file.

The photography, preferably the film or glass negatives exposed on site, will be positioned on the platens in the photogrammetric instrument, again avoiding all finger contact with the emulsion in order to avoid any physical deterioration, and the orientation procedures carried out as previously described.

Figure 3.13 shows a typical analytical photogrammetric instrument as manufactured by Galileo Siscam of Italy. The two film negatives can be seen positioned on the glass platen. Figure 3.14 shows the Wild Leitz BC3 instrument. Other instruments of a similar design and usage include those manufactured by Zeiss, Intergraph and Kern (now part of the Wild Leitz group).

It is a standard approach in conventional surveying to observe two or three rounds of angles in order to increase the precision of the operation. Making a series of readings and taking the mean of results will always strengthen the solution, and help to highlight any poor observations. The same is true in photogrammetry, and it is accepted good practice for the photogrammetrist

Figure 3.13 Galileo Siscam Digicart 40. This analytical instrument is driven by a '386 PC' and relies on a high-resolution graphics screen to portray the recorded data rather than the conventional plotting table. (Photograph courtesy of AMC Ltd.)

to extract multiple readings for each point being recorded. The means of sets of coordinates values can then be calculated across all three dimensions to produce the final results. However, it is also useful to indicate the spread of the recorded values for each point, as any poorly observed readings will result

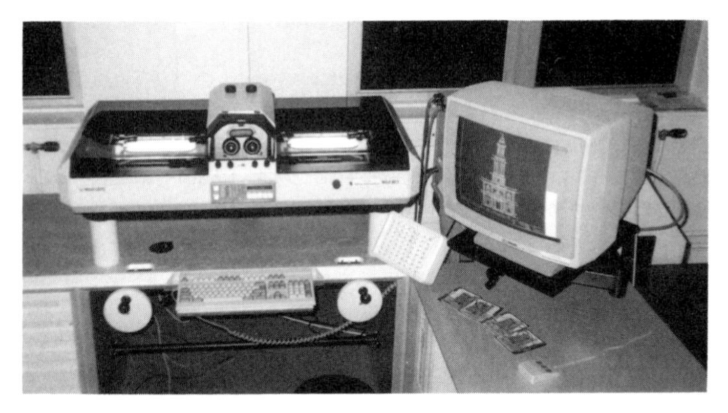

Figure 3.14 Wild Leitz BC3. Apart from subtle differences in construction and operation this and other similar instruments are all capable of being used in monitoring exercises. This particular instrument is driven by handwheels instead of the freehand devices as used by Galileo and Intergraph. (Photograph courtesy of Survey International.)

in a large spread, allowing the opportunity for reobservation if this is deemed to be necessary.

3.4.7 Data processing and output

Normally the main form of output for photogrammetric monitoring projects is a tabulated set of coordinate values. The only data processing required in these circumstances is to provide the correct data transfer format. However, this form of data may well, as previously mentioned, be supplemented by a graphical output, such as one of the following:

Sections: by overplotting sections or profiles from two different epochs, possibly with an exaggerated horizontal scale, any movement becomes highlighted.

Contours: by extracting the differences in movement in any one axis for each point measured, it is possible to produce a contour plot which gives an overall view of both the magnitude and direction of movement. The optimum contour interval needs a judgement based on the maximum and minimum deflections.

Digital surface model: such a model consists of a number of points recorded either in a set pattern or as judged by the photogrammetrist to best represent the surface of the building. A digital surface model will normally be supplied as a series of three-dimensional coordinates in computer-compatible form, possibly for one of the many CAD systems available.

Although there is still no clearly defined or nationally accepted digital exchange format, most end computer systems can take data in one of the more common graphic exchange formats and these are supported by most photogrammetric software packages.

3.5 Project planning

So far no mention has been made of the accuracies that can be achieved by photogrammetry. This is because there are so many factors affecting the performance of photogrammetry and these factors require careful consideration before any statements on accuracy can be made.

When considering a new project the photogrammetric engineer in charge will be faced with two options—either to devise a scheme to achieve a stipulated accuracy or to make a finite statement as to what accuracy can and will be achieved. In either situation it is best to carry out the project planning with a realistic sense of pessimism.

For any monitoring project a site visit is essential, as no matter how many plans and photographs are examined a true 'feel' for the site cannot be obtained without some first-hand knowledge. The early planning stages are time-consuming but essential on the part of both the project engineer and the photogrammetric engineer.

The first discussion point should always be to ascertain the purpose of the project and the likely performance of the building in terms of movement. From this information it will be possible to determine the time periods (epochs) between measurements.

One very important factor concerning the performance of the building is the amount of sway in the structure as a result of wind. If sway is significant then it creates a major problem in relating the data to an absolute datum and the measurement may need to be restricted to local or relative movement only. Having said that, it is an interesting point that photogrammetry is an excellent tool for the measurement of sway because of the ability to freeze the motion of the building at its limits of movement.

Having understood the character of the building the next stage concerns the practical positioning of the camera and control stations in order to test the proposed geometry for accuracy.

The analytical instrument will produce information during the orientation stages concerning the coordinates of the control data and their differences against the photogrammetric observations. These residual values allow the photogrammetrist to override the control values by differential weighting, thus, in effect, correcting some of the error in the control data. However, at the planning stage of a project, it is safer to add together the theoretical accuracies for both the photogrammetry and the control, and then add the required pessimism factor into the equation.

In order to compute a guide figure for the accuracy of the control by intersection techniques, the following formula should perform satisfactorily when the theodolite intersection geometry is close to the ideal situation:

$$\text{Accuracy (mm)} = \frac{[\text{Theodolite to control distance (m)}]^2}{\text{Theodolite base (m)} \times 50} \times \text{Instrument accuracy (seconds of arc)}$$

The guide figure for the photogrammetric accuracies can be calculated using a similar formula.

$$\text{Accuracy (mm)} = \frac{[\text{Camera to object distance (m)}]^2}{\text{Focal length (mm)} \times \text{Camera base (m)}} \times \text{Instrument accuracy (micrometres)}$$

These two formulae are designed to give a realistic indication of expected accuracies and are based on experience of actual accuracies achieved. In all cases the worst possible case should be assumed, i.e. the longest camera to object distance. However, as with any formulae of this nature allowances must be made for the less obvious.

It will be seen that, using a nominal 6 μm (6×10^{-6} m) for the instrument accuracy, a figure of 2.4 mm is obtained for photogrammetric accuracy when camera to object distance and base are both 40 metres (100 mm focal length). By doubling the camera to object distance, the camera base needs to be multiplied by a factor of 4 to obtain the same accuracy levels, but will the photogrammetrist be able to position the measuring marks to the required level of accuracy with a much smaller photoscale and an increased obliqueness in the photography?

Whilst the equation compensates for increased scale and the observational decline in pointing accuracy, the experienced photogrammetric engineer on site needs to be able to apply his own subjective factor to the formula. This is done by increasing the measuring accuracy of the instrument involved by a suitable factor, of possibly 2 or more.

In practical terms, the best intersection angle for both control and photogrammetry is 90°. This gives the optimum 'cut' and is thus preferred. If unlimited access to the area in front of the facade to be measured is available then the optimum camera stations can be sited so as to achieve the 90° intersection based on the centre of the facade. If possible the two camera stations (or more if necessary) should be equally sited from the centre of the facade, in order to create balanced geometry. The distance back from the facade will often be dictated by the height of the building as a result of buildings often being higher than they are wide. Again, the use of analytical techniques permit the cameras to be angled not only towards each other (convergence) but also upwards (tilt) to accommodate tall structures. However, it is necessary to take into account the diminishing levels of accuracy to be expected when tilting the cameras as a result of the increased camera to object distances spoiling the geometry.

If the effect of tilting the cameras decreases the accuracy by more than is acceptable, then it will be necessary to introduce the use of elevated camera stations, either hydraulic access platforms or scaffolding towers. Unstable camera platforms should only be used as a last resort and not at all unless available light is sufficient, or fast film emulsions are available to counteract the effects of any camera movement.

The angle of view of the metric cameras previously mentioned are approximately 66° × 84° for the Zeiss and 54° × 66° for the Wild, albeit that the offset format complicates the latter figure. When calculating the preferred camera stations and angles of convergence it is useful to make up a simple template with the principal axes and angles of view drawn thereon. Providing this template is of sufficient size it can be used with any scale of site plan and soon becomes an invaluable device.

With regard to planning the control stations, this follows similar criteria to the positioning of the camera stations, i.e. if the geometry is right for photogrammetry then a similar pattern will be correct for the control observations. However, with control sometimes being outside the monitoring area, the control stations may also need to be positioned outside the camera station to achieve a similar geometry. The positioning of the control points needs detailed investigation. If the building is stable and not subjected to any significant sway (see Merritt Point example), then control can be positioned within the facade but away from any areas of likely movement. However, if the whole facade is subject to movement, then as much control as necessary needs to be sited on stable surfaces, perhaps adjacent buildings unaffected by movement, and on the ground. In exceptional circumstances rigid scaffolding towers can be erected to offer a stable surface on which to position the control points.

However, as previously discussed, if relative monitoring is acceptable, then photogrammetry can achieve the results by employing a bundle adjustment solution. In all cases, the control targets need to be positioned around the limits of the photographic cover, with some additional points included in the centre of the model, if practical. Obviously, if the structure is dynamic because of wind sway or other causes, then only those points on stable surfaces need be recorded, and thus only relative movement can be measured at the photogrammetric data capture stage.

3.6 Case histories

3.6.1 Criteria for using photogrammetry

The question as to whether or not photogrammetry is the most suitable option for any monitoring project has to be answered at the initial planning stage as the criteria are governed totally by the site, with camera station positioning being the most critical factor.

It will be appreciated by now that no definitive answer exists as to the suitability of photogrammetry in general terms, as each building and site is unique. However, to assist the reader in judging the usefulness of photogrammetry two documented examples of photogrammetric monitoring follow.

3.6.2 Smithfield cold store (53 Charterhouse)

The first case history describes a project undertaken by AMC Ltd of Pewsey, Wiltshire, England, during the winter of 1989/90. AMC specialise in architectural photogrammetry, mainly for conservation purposes, and have been involved in many projects including a number of fire-damaged building surveys, such as at Uppark House on behalf of the National Trust (Architects: The Conservation Practice) (see Figures 3.2–3.5).

The Smithfield cold store is a Victorian building with a main elevation fronting onto Charterhouse Street in London and measuring 32 m wide by 22 m high. During the late 1980s, a fire destroyed the majority of the interior and, because of the 0.5 m depth of cork lining on the internal walls, burnt with a great intensity, causing serious structural faults.

With the building previously having been listed, it was decided to consider the retention of as much of the building as possible, and in particular the front elevation. Photogrammetry was chosen by the consultants (W.S. Atkins & Partners) to record the detail of the main elevation to a scale of 1/50. However, owing to the structural deformations that occurred within the facade, it was also decided to produce a three-dimensional surface model. The production of this model was to make use of the three-dimensional facility that photogrammetry offers, and thus it was proposed that both aspects of the project would be carried out from the same raw data, i.e. the photography.

The criteria set for the project included an accuracy in all three dimensions of ± 3 mm, and that, if required at a later date, the facade could be rephotographed to determine if any changes had occurred in the more deformed areas.

A site visit revealed that the main areas of the facade for monitoring had sufficient surface texture (glazed brickwork with fine mortar jointing) to avoid the need for targeting. This was important as in this surface-modelling exercise a relatively large number of points were to be recorded. Although in the first instance the computer-interpolated contours were compared, rather than the individual measured points, should a repeat run occur in the future the individual point coordinates would then become critical.

Had the building surface been smooth, then stereoscopic fusion would have been weak, and the areas would have had to have been targeted or, alternatively, the surface given added texture by random paint spraying or similar methods.

As previously mentioned, the modelling of the building surface was to complement a conventional elevational survey and it was, therefore, preferable to ensure that the photography taken was suitable for both tasks, ruling out the use of the convergent photography on this occasion.

The elevation drawings were at a nominal scale of 1/50 (nominal because the data were supplied in CAD-compatible format), and a photoscale suitable for this aspect would need to be in the region of 1/200, which, using a camera with a focal length of 100 mm, relates to a camera to object distance of 20 metres. It was possible to gain access to the parapet level of the building opposite (see Figure 3.15) and this permitted not only adequate photography but also acceptable geometry. This was taking into account the fact that good 'parallel' photography was essential to achieve an ideal stereoscopic model for optimum viewing comfort. It can be argued that when parallel photography is involved, the presence of an image which is comfortable to view in 'stereo' is directly related to achieving optimum precision in a non-targeted situation.

The main facade of the cold store was photographed with a Zeiss UMK 10/1318 metric camera, with a focal length lens of 100 mm, using Ilford FP4

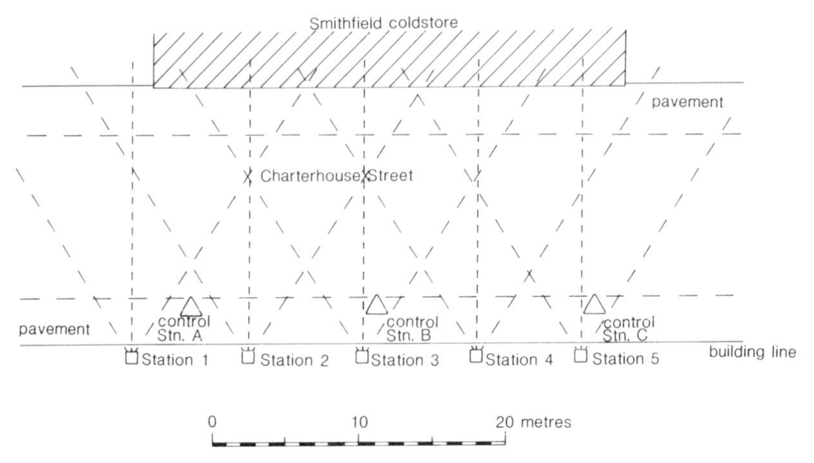

Figure 3.15 53 Charterhouse Street site plan. The geometry was designed to make use of the available access from the parapet opposite. An object distance to camera base ratio of just over 2 : 1 was achieved, at the same time retaining parallel camera axes.

emulsion (see Figure 3.16). The maximum camera to object distance was 18 m, and the average camera base was 7.8 m. Control was to targeted points from a permanently marked traverse using a Geotronics 440 total station to measure both the bearings and the distances between the traverse stations.

By applying the formula given in section 3.5 a guide figure for the accuracy of ± 2.5 mm was obtained for the photogrammetry using an instrument accuracy of 6 μm. As only local movements were being measured, the control accuracy was calculated to be ± 0.5 mm, giving a total figure of 3 mm, which met the requirements of the structural engineers.

The facade was recorded using one of AMC's Digicart 40s and supplied as an AutoCAD drawing file (see Figure 3.17). The grid of points was recorded at 1-m centres and subsequently fed into an Intergraph system. Contours were then interpolated at 20-mm horizontal intervals showing a marked pattern of distortion in the upper parts of the elevation.

This project represents a very straightforward type of photogrammetric monitoring. Although the exercise has not yet been repeated, the initial parameters were determined with repeat runs anticipated to measure the expected movement when structural integrity is diminished by partial demolition.

3.6.3 Merritt Point fire tests

This case history involves a project at the other end of the spectrum to the Smithfield cold store exercise. Indeed, the only common denominator is the fact that fire is once again involved, albeit this time in a controlled manner.

Figure 3.16 53 Charterhouse Street (Smithfield cold store), one of the photographs (camera station 3) showing cracking in the brickwork caused by the fire. (Reproduced with the permission of W.S. Atkins & Partners.)

In 1986 the Building Research Establishment commissioned Wimpey Laboratories to undertake a series of three fire tests on a 22-storey block of vacated domestic apartments within the London Borough of Newham. The building was called Merritt Point; its immediate neighbour and sister was Ronan Point, which at the time of the fire tests was being dismantled.

The tests were to take place on the third, fifth and sixth floors, and photogrammetry was introduced to record any movement in the flank wall for the first two tests, i.e. those on the fifth and sixth floors.

Merritt Point was built with exterior walls consisting of precast concrete panels connected directly to the concrete floor slabs, which were also precast. This produced a structure consisting, in effect, of a series of boxes stacked one upon the other. The structural strength of these blocks of apartments became suspect following the partial collapse of Ronan Point. This resulted

Figure 3.17 53 Charterhouse Street. Part of the elevation drawing, plotted from an AutoCAD file, showing substantial cracking in the brickwork. (Reproduced with the permission of W.S. Atkins & Partners.)

in substantial structural strengthening being carried out on all the similar buildings, including Merritt Point. The aim of the fire tests was to help establish what structural deformations would occur in this type of building during a typical domestic fire situation.

The tests were centred in the lounge area of the apartments, which was one room removed from the flank wall. However, as the floor slabs were expected to distort and expand during the fire, it was rightly assumed that the flank wall would move with it. Whilst it was relatively straightforward to measure movement internally by the use of more conventional direct or contact techniques, the exterior was, to all intents, inaccessible, and thus the application of photogrammetry was examined.

For each test a total of 99 points were monitored, although this was extended to 132 points for the second test. These numbers equate to 33 points per floor, with the fire floor plus one floor up and one floor down (or two up and one down in the latter case) comprising the test area. The 33 points per floor level were distributed over the three concrete panels forming the left-hand half of the flank wall. Figure 3.18 shows the distribution of these points.

During each fire test the timings of the recordings were governed by the category of the fire. In the first test, the room was heated by gas burners with temperature increments of 200°C, 400°C and 600°C. At each increment, time was allowed for the heat to be absorbed into the structure before the recording took place. This was then followed by the next burn to raise the temperature up to the next level. Recordings were made prior to the start of the first burn to produce the base data, at each of the three increments, and a final one approximately 30 minutes after the burners were extinguished. This pattern of recordings produced measurements throughout the main period of movement within the structure.

The second test was different inasmuch as the gas burners heated the room to 1000°C and then held the temperature at this level for a period of 30 minutes. Recordings were required before the burn began, upon reaching 1000°C, and at 10-minute intervals until the gas burners were extinguished. As with the first test, a final recording was made half an hour after the extinguishing of the burners when some retraction would have occurred.

During the partial collapse of Ronan Point, debris spread up to 20 metres away from the base of the building. There was a very remote risk that one or more flank wall panels or other parts of the structure might break away during the tests, and therefore a ban on personnel within a 25-metre radius of the test zone was enforced.

With regard to precision levels, for the results to be of use an accuracy of ±1 mm was requested, but this was later adjusted to ±2 mm owing to the introduction of the 25-metre prohibition zone.

The project criteria were therefore very well defined. The site itself, as far as access was concerned, caused no major problems, although some minor

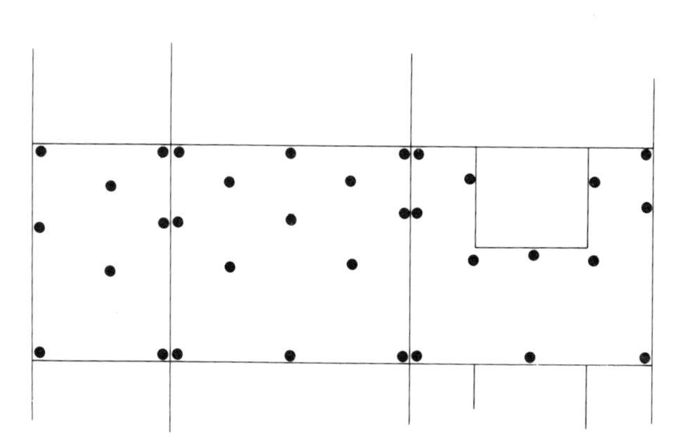

Figure 3.18 Merritt Point, sixth floor. Distribution of targets for monitoring.

inconvenience was caused by it being split into two levels, with the right-hand part of the site some 3 m higher than the left-hand portion.

It is useful at this point to consider the justifications for using photogrammetry on this project, as they fulfil three of the originally discussed criteria (section 3.3):

(1) Structure dynamic, with movement of up to 12 mm expected.
(2) Multiple readings involved with a relatively large number of points involved each time.
(3) Remote sensing essential because of the possible dangers of falling debris.

Figure 3.19 shows a plan of the site with the positions of the camera and control stations. The geometry indicated on this plan was calculated following a thorough site visit accompanied by the engineer in charge of the tests.

Owing to the shape of the elevation it was decided to use a pair of Wild P31 cameras, the offset format helping to reduce the convergence of the principal axes to a minimum. The nature of the project called for local changes to be recorded, thus allowing for the control to be positioned within the elevation, although as far away from the fire zone as possible. It was therefore decided to capture the whole of the width of the building up to the 15th-floor level. As it happened, because of the nature of construction, there was minimal sway in the building, greatly increasing the precision of the control observations and computation.

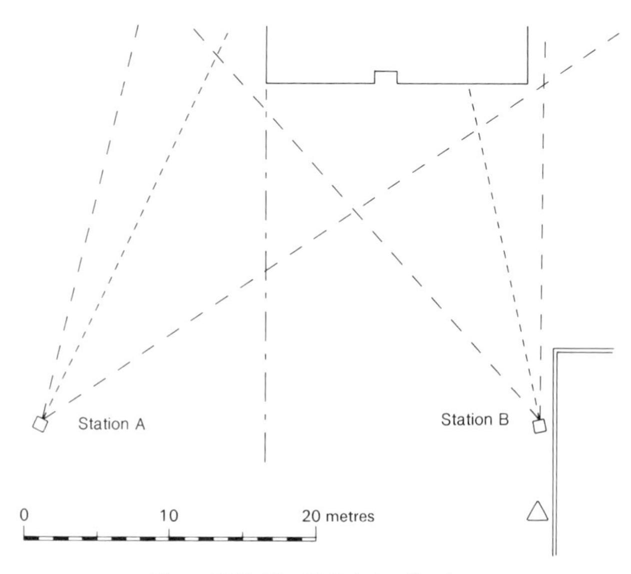

Figure 3.19 Merritt Point—site plan.

The left-hand camera station (A) was situated at a height above datum of 6 metres, some 3 m below the level of the right-hand station (B). The siting of the cameras created a maximum camera to object distance of 36 m. With a camera base of 30 m and a nominal focal length of 100 mm for the P31 cameras, the only other factor to determine, in order to calculate the likely accuracy levels, was the photogrammetric instrument accuracy. The data extraction was to be carried out by BKS Surveys Ltd on their Wild BCI analytical photogrammetric instrument, which was calibrated to within 2 μm. It was decided that, as the observations were to be to well-defined targets, a final instrument accuracy of 4 μm was appropriate.

By applying the formula previously given in section 3.5, an accuracy of ± 1.7 mm was expected, which combined with a control accuracy in the region of ± 0.5 mm produced a total of ± 2.2 mm, which was only just outside the requested accuracy. As this calculated accuracy was for the worst case (i.e. the longest camera to object distance), a general level of accuracy within ± 2 mm was expected. Indeed, when considering the best case, the expected photogrammetric accuracy reduces to ± 1.2 mm.

Figure 3.20 shows a typical pair of photographs from the Merritt Point project. It will be noticed that there appears to be more rotation to the left-hand image than to the right-hand image. The reason for this is that the geometry was based solely on the area to be monitored, i.e. the left half of the flank wall, whilst the cameras were rotated to capture the whole width of

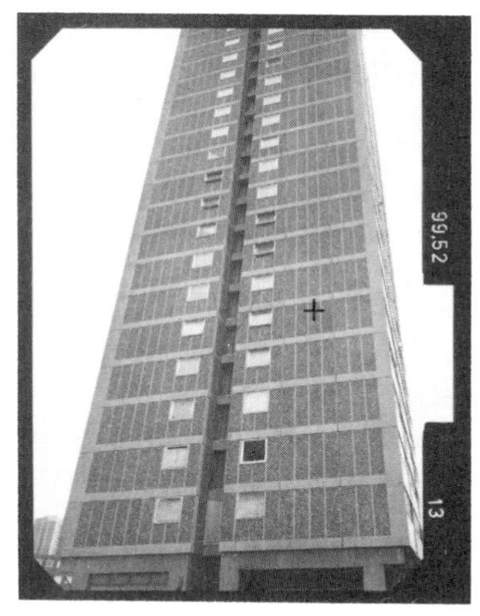

Figure 3.20 Merritt Point. A pair of simultaneously exposed photographs taken using two Wild P31 cameras. These cameras have a 5″ × 4″ format (125 mm × 100 mm) with an offset format. The principal points have been highlighted to demonstrate the offset optical centre. The figures 99.08 and 99.53 are the calibrated focal length values in millimetres. (Reproduced with the permission of the Building Research Establishment; consultants, Wimpey Laboratories Ltd.)

the building, thereby increasing the rotation to the left-hand camera. This situation was exacerbated by the demolition of Ronan Point, as it transpired that the optimum position for the left-hand camera could not be achieved because of debris blocking the view. In theory, the increase in the camera base improved the achievable accuracy, although in practice by a negligible amount.

The right-hand camera station was positioned as tight as possible against the wall shown on the plan and ideally the right-hand control station would have been further to the right, as the control on the right-hand side of the building was as important as the left.

Knowing that the effects of the fire tests were likely to extend up to floor 8, and down to floor 2, it was decided to position control targets across the ground-floor level, and across the ninth-, twelfth- and fifteenth-floor levels with an additional point on the fifth floor on the far right-hand side (see Figure 3.21). The control targets were the same design as the monitoring targets, a black and white bull's-eye, 25 mm in diameter, printed onto a semi-rigid heatproof plastic material. The targets were fixed in place using an epoxy resin, the operators working from a cherry-picker to obtain access. Above the seventh floor the targets were positioned by leaning out of the appropriate windows.

The control was observed using two Kern E2 1-second electronic theodolites linked to a Husky Hunter portable computer. The accuracy indicator checks (dZ values) were less than 0.5 mm up to the ninth floor and 1.0 mm up to the twelfth floor. The fifteenth-floor values were around 2.0 mm, but this was expected. The accuracy of the control proved to be compatible with the requirements, and excellent residuals were later experienced when orientating the pairs of photographs in the photogrammetric instrument.

For the photography, Ilford FP4 cut film was used, downrated to 64 ASA and processed in Perceptol. The P31 lens was set to a maximum aperture of f8 to permit optimum film resolution. Depth of field was adequate at this aperture to cover the focusing range required. Synchronisation of the camera shutters was by radio communication.

In order to verify the integrity of both the control and the results obtained from the photogrammetric recording, the surveyors reobserved the control targets and a selection of points, both during and at the end of each test. The reobservation of the control confirmed the indicated accuracies mentioned. The comparison with the photogrammetrically produced data is discussed later.

The photogrammetric data capture stage commenced with the collation of the control data and the weighting of the values to achieve a logical solution based on the dZ values supplied. Each pair of photographs was orientated in the instrument with the resulting residual values compatible with the weighted values. Owing to the positioning of the control, it was possible to carry out the relative and absolute orientations as a single-step adjustment.

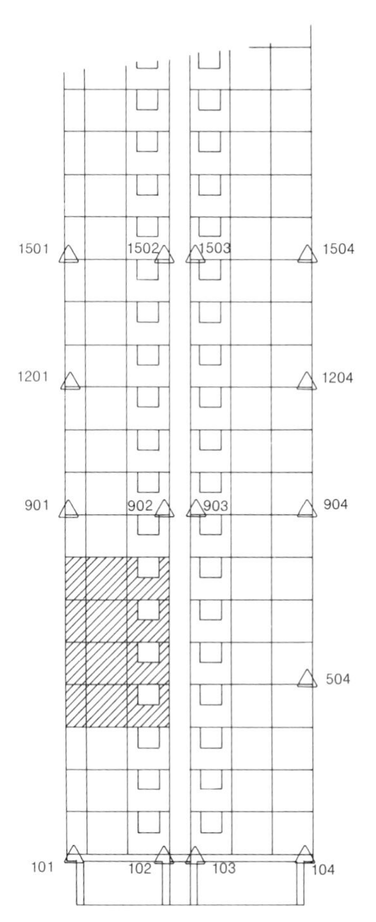

Figure 3.21 Merritt Point. Distribution of control targets. The hatched area represents the area for monitoring during the fire tests.

Prior to the targeted monitoring points being observed and recorded a random selection of the points were observed and reobserved to test the repeatable accuracy. This test confirmed that the repeatable accuracy was of the order of 1 mm within the test area. For the actual data abstraction, each point was read three times and the mean of results calculated.

Table 3.1 is an example of the values recorded by photogrammetry during the second and more dramatic of the tests. Point 601, nearest to the fire on the edge of the floor slab, moved a total of 24 mm, the maximum movement experienced.

From Table 3.1 it can be seen that the x and z values (ground coordinate system) change very little, and within the calculated accuracies. However, the

Table 3.1 Merritt Point—point 601 movement values in metres.

Test no.	x	y	z	y (mm)
2/1	8.610	9.882	22.896	0
2/2	8.611	9.880	22.894	2
2/3	8.610	9.869	22.894	13
2/4	8.612	9.862	22.895	20
2/5	8.610	9.858	22.894	24
2/6	8.611	9.863	22.895	19

y values show that an indicated movement of 24 mm occurred shortly before the burners were finally extinguished after the 30 minutes of the test.

When comparing the photogrammetric results with the theodolite checks, the figures agreed to within 1 mm in virtually all cases.

Inside Merritt Point a number of other pieces of monitoring apparatus were in operation during the tests, including a laser which was set up to record movement in the floor slabs. The values recorded by this device positioned on the inside of the flank wall mimicked the values on the outside recorded by photogrammetry also to within 1 mm. A more general analysis of the results indicated that the required accuracy of ± 2 mm had been achieved for over 97% of the results, with the remainder within ± 3 mm.

On a project like the Merritt Point it would be virtually impossible to undertake the measurement by any other method. Although relative movement was to be monitored, the data recorded could be regarded as absolute because of the rigidity of the structure. However, if the building had been liable to sway, then relative movement could still have been achieved. Output on this project consisted of a tabulated coordinate listing plus contour plots showing the maximum movement of the flank wall slabs during each test, the contours being at 5 mm horizontal intervals.

3.7 Conclusions

Photogrammetry offers a flexible monitoring technique with distinct and unique advantages (dynamic situations, mass data collection, remote sensing, etc.). There is also the value-added aspect of reappraisal using the photographic records.

Obviously the large capital value of the photographic, photogrammetric and control equipment will prohibit all but the specialist organisations in carrying out this type of work, but with the level of expertise needed the use of properly equipped and experienced specialist contractors must be recommended.

Appendix: Moiré photography

The principle of moiré has been applied by Burch and Forno (1982) at the National Physical Laboratory to the measurement of in-plane displacements at high sensitivity on buildings and other engineering structures.

Moiré is the well-known effect produced when two regular patterns are superimposed. In their technique, high-resolution moiré photography, one pattern is attached to the surface of the structure and a photograph is then taken in its initial, undeformed, state. By comparing this negative with another taken at a different state, a moiré fringe pattern is generated. This is a direct representation of the displacements where each fringe corresponds to a movement, relative to an adjacent fringe, of one pitch of the pattern. Figure 3.A1 shows an application to a timber joint in which the displacement interval is 50 μm. Automatic analysis of the fringe pattern can generate a strain contour map which will allow identification of areas of importance.

A standard camera with a high-quality lens is suitable for the process. It has to be modified by the insertion of a slotted mask in the aperture in order to tune the imaging response so as to achieve the necessary sensitivity. A variety of methods may be used to target the subject, but the most convenient for structural application is paper printed with an array of dots, typically at the rate of 1 dot per millimetre, as described by Forno (1988). The paper is stuck to the surface to be studied as shown in Figure 3.A2, in this case involving a very detailed study of a facade subject to subsidence. Here, a 15 m wide elevation was photographed from a distance of 15 m to provide a displacement resolution of 0.2 mm.

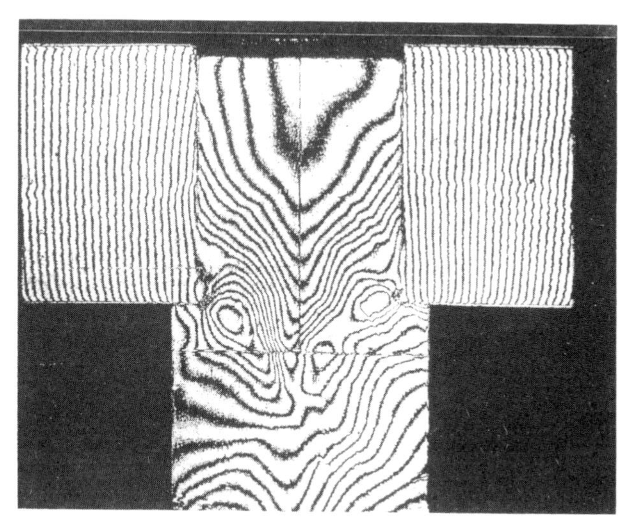

Figure 3.A1

Figure 3.A2

Clearly, this form of targeting may restrict applications of the method, for reasons of appearance, but where the examination of a complete field of displacement is required and where the substantial variations of strain are expected, the only comparable measurement technique is photogrammetry.

Acknowledgements

The author would like to extend his thanks to the following for their permission to use their project material. Uppark House: The National Trust and The Conservation Practice; 53 Charterhouse: W.S. Atkins and Partners and AMC Ltd; Merritt Point: The Building Research Establishment, Wimpey Laboratories, BKS Surveys Ltd and the London Borough of Newham.

The author would also like to thank Hugh Anderson ARICS for his advice and technical assistance.

Colin Forno, Division of Mechanical and Optical Metrology, National Physical Laboratory, Teddington, England is gratefully acknowledged for the Appendix to this chapter.

References

Anderson, H. (1985) The first photographic negative. *Photogrammetric Record* **XI**(66), 641–644.

Atkinson, K.B. (1980) Vivian Thompson (1880–1917): Not only an Officer of the Royal Engineers, Appendix. *Photogrammetric Record* **X**(55), 36–37.

Brown, D.C. (1982) STARS, a turnkey system for close range photogrammetry. *International Archives of Photogrammetry* **24**(V/I), 68–89.

Burch, J.M and Forno, C. (1982) High resolution moiré photography. *Optical Engineering* **21** (4), 602–614.
Carbonnell, M. (1969) The history and the present situation of the application of photogrammetry to architecture. *Proceedings of the conference on the Application of Photogrammetry to Historic Monuments, Saint Mande, France*, 1968. ICOMOS, Paris.
Dallas, R.W.A. (1980) Architectural and archaeological recording, In *Developments in Close Range Photogrammetry*—1, ed. K.B. Atkinson, Applied Science Publishers, Amsterdam, pp. 81–116.
Foramitti, H. (1966) Photogrammetry in the hands of the building expert. *Deutsche Bauzeitung* **9**, 786–92, and **10**, 874–80.
Forno, C. (1988) Deformation measurement using high resolution moiré photography. *Optics and Lasers in Engineering* **8**, 189–212.
McDowall, R.W. (1972) Uses of photogrammetry in the study of buildings. *Photogrammetric Record* **VII**(40), 390–404.
Thompson, E.H. (1962) Photogrammetry in the restoration of Castle Howard. *Photogrammetric Record* **IV**(20), 94–119.

4 Dynamic testing

B.R. ELLIS

4.1 Introduction

This chapter deals with the dynamic testing of structures or structural elements. The data which are obtained from the tests can be used either to check the accuracy of mathematical models of the structure or to predict how the structure will behave under given load conditions.

There are numerous different test methods which can be used to determine the characteristics of a structure, and in general the more information required, the more complex and time-consuming the test. However, a lot can be learnt from a few simple tests providing the results are interpreted correctly and the limitations of the test methods are understood. The objectives of this chapter are to explain:

(1) What can be measured.
(2) How it can be measured.
(3) How the results can be interpreted and used.

The next section provides a brief description of what can be measured, followed by a detailed section discussing how to measure various structural characteristics, dealing initially with the simpler measurements. Subsequent sections deal with instrumentation, use of information, non-linear behaviour and integrity monitoring. An appendix is provided which presents the mathematical conversion from Cartesian to modal coordinates. Throughout the chapter examples are given from tests with which the author has been involved, to show the type of data which are likely to be encountered.

This chapter does not deal with monitoring building response to earthquakes, ground-borne vibration or machine vibrations.

4.2 What can be measured

One of the first equations that will be encountered in texts on structural dynamics is the equation describing the single degree of freedom viscoelastic model, that is:

$$m\ddot{x}+c\dot{x}+kx = p(t) \tag{1}$$

where m is the mass, c is the damping, k is the stiffness, $p(t)$ is the force with time, $\ddot{x}$ is the acceleration, $\dot{x}$ is the velocity and x is the displacement.

When the behaviour of buildings or building elements is examined it is found that it can be characterised by the combination of a number of independent modes of vibration each of which can be described by an equation similar to eqn (1). Moreover, for a number of applications, only the fundamental mode of vibration (the lowest frequency mode) needs to be considered, hence the simple one degree of freedom model can be used to describe how the structure behaves.

It should be noted that when eqn (1) is used to describe a mode of vibration, the parameters are all modal parameters and the coordinates are modal coordinates, and are not the conventional terms used in static analysis and Cartesian geometry. However, the two are mathematically related, and this relationship is described in the appendix to this chapter. This mathematical manipulation is only necessary in the more comprehensive test schemes.

The single degree of freedom equation can also be manipulated to yield a useful approximation which is reasonably accurate for the damping levels encountered in buildings:

$$\omega \simeq \sqrt{\frac{k}{m}}$$

where ω is natural circular frequency.

This is related to the natural frequency f and its inverse, the period T, by

$$\frac{\omega}{2\pi} = f = \frac{1}{T}$$

The damping coefficient c is related to the damping ratio ζ by the following equation:

$$c = 2\zeta\omega m$$

Substituting these expressions into eqn (1) and dividing by m gives:

$$\ddot{x} + 2\zeta\omega\dot{x} + \omega^2 x = \frac{p(t)}{m}$$

To describe the terms graphically and explain what they mean in simple terms, it is worthwhile repeating a diagram shown in most texts on structural dynamics which is the response of a single degree of freedom model to a constant force over a range of frequencies (Figure 4.1).

The curve is defined by three parameters:

(1) The natural (or resonance) frequency which for systems with low damping, i.e. most practical systems, can be considered to be the frequency of the maximum response.

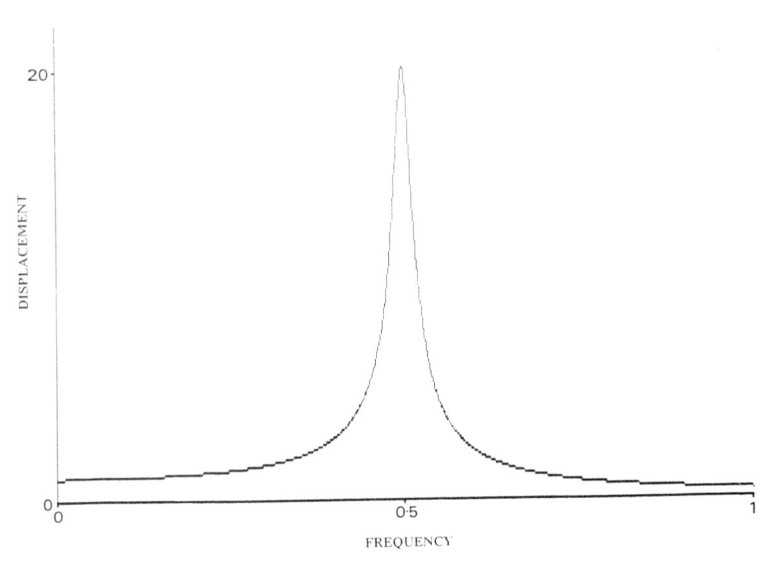

Figure 4.1 Response of a single degree of freedom model to a constant force.

(2) The stiffness, which relates the static force to static displacement, i.e. displacement at zero frequency.
(3) The damping, which defines the shape of the curve: the lower the damping, the narrower the resonance curve and the higher the amplitude at resonance.

The curve given in Figure 4.1 has been generated for a system with a 2.5% damping or $\zeta = 0.025$, a stiffness of 1 and a frequency of 0.5 Hz, and is given to show a typical value which might be found in buildings. A term which is often encountered is magnification factor, which relates the response at resonance to the static response and is equal to $1/2\ \zeta$. For this example the magnification factor is 20.

The following section deals with the measurement of frequency, damping, mode shape and stiffness, thus providing experimental measurements to define the structural system.

4.3 Types of test

4.3.1 Parameters which can be measured

In order to describe the types of test which can be used, consider the problem of defining the fundamental mode of vibration of a structure. Assume that the mode does not exhibit non-linear characteristics, i.e. its characteristics do not change with changing load or amplitude of vibration, hence the mode can be uniquely defined by the following four modal parameters:

(1) Natural frequency (or resonance frequency).
(2) Damping.
(3) Stiffness.
(4) Mode shape.

In general it is easy to measure the frequency, more difficult to measure damping, and measurements of stiffness and mode shape require tests using specialist instrumentation. However, for many applications only the natural frequency is required, and it is the measurement of natural frequency which is discussed first.

4.3.2 Frequency from a simple impact test

In order to illustrate the use of impact tests to measure frequency it is useful to examine a straightforward example. Assume the fundamental frequency of the vertical mode of vibration of a floor is required. The basic equipment necessary to take a simple measurement is a transducer (say a geophone) which will measure motion in a vertical direction and a means of displaying the signal, a y–t plotter or an oscilloscope. Place the transducer in the centre of the floor and secure it with some tape, then display the output of the transducer on the oscilloscope. The fundamental mode of the floor can then be excited by hitting the centre of the floor with a compliant mass (a person jumping once). The time base, amplitude scales and triggering of the oscilloscope will need to be adjusted to give a reasonable display such as that given in Figure 4.2. This figure, which is a measurement of the response of the floor in the author's office, shows that the impact excited one mode of vibration.

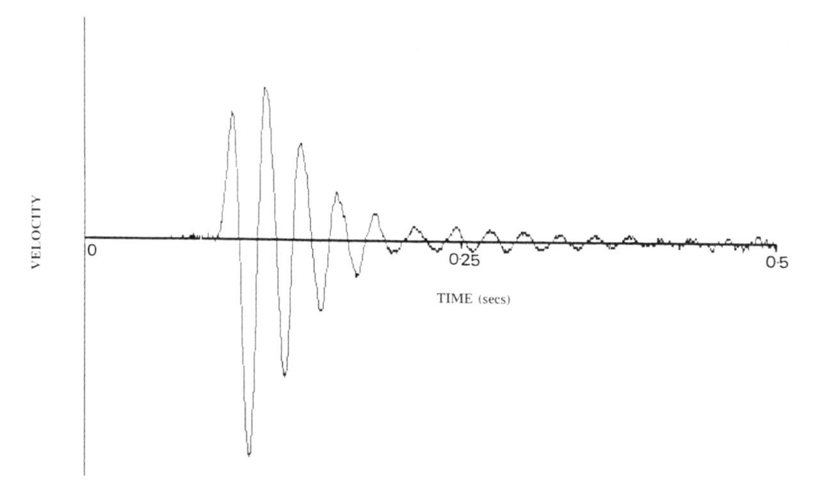

Figure 4.2 Response of an office floor to an impact.

In some cases two or more modes might be seen, although the higher frequency ones usually decay rapidly to leave the response of the fundamental mode, which is a decaying sinusoid. From this the time taken for each cycle (the period) can be calculated and therefore the frequency can also be calculated, i.e. the inverse of the period. Obviously being able to take a copy of the trace means that the frequency can be calculated more accurately, but a reading on an oscilloscope may be perfectly adequate. Although this is a simple example, the basic principles still apply if measurements are required on, say, a wall or a complete building, although the impacting mechanism may need to be changed. In many cases this fundamental frequency can be compared with calculations to provide feedback to check design assumptions.

If the impact on the floor is recorded and available for analysis on a computer then more accurate analysis techniques can be used. For example, the data can be transformed using a fast Fourier transform (FFT) procedure to show a plot of response against frequency (see Figure 4.3), and in this case it can be seen that only one mode has been excited. This form of analysis, which requires some specialist equipment, is dealt with later.

It may be of interest to the reader to estimate as accurately as possible the frequency from the decay in Figure 4.2.

4.3.3 Frequency from ambient response

With tall buildings, it can be appreciated that providing an impact to excite the fundamental mode of the whole building may not be easy, although

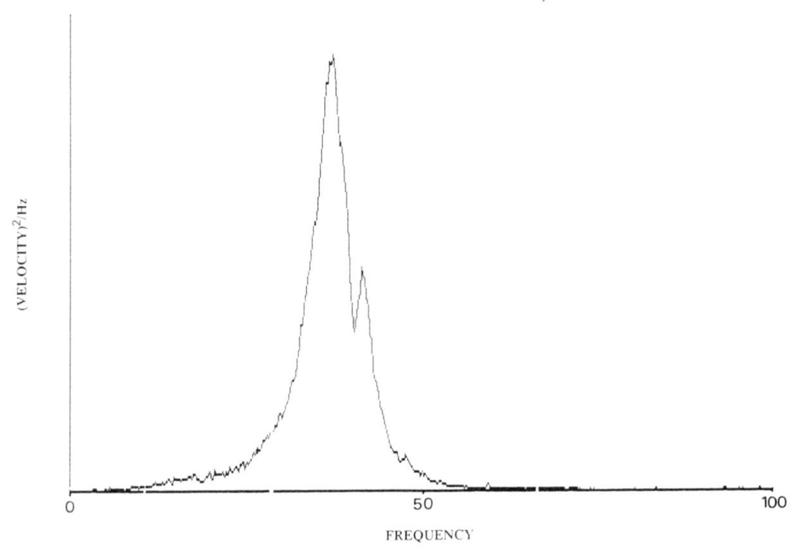

Figure 4.3 Autospectrum of response shown in Figure 4.2.

various experiments have been tried, for example rockets attached to the building by rope have been fired from the building, or wires have been attached to the top of the building and a sideways load applied and then suddenly released. However, a building is often subjected to wind loading, and this will excite the fundamental mode.

Taking the basic equipment of a transducer (this time one to measure horizontal motion) and an oscilloscope it is possible to estimate the fundamental frequency of a building provided that a reasonable wind is blowing. Align the transducer with the axis of the building which is of interest (fundamental modes will be found for two orthoganol translation directions and the torsion direction). If data on a translational mode are required, place the transducer near to the geometric centre of the building to minimise the torsional response. The oscilloscope will show what appears to be low-level random motion, but every so often sinusoidal motion will occur which will be motion in the fundamental mode and hence the frequency of the sinusoid can be identified (see Figure 4.4). Occasionally higher order modes can be seen, but usually the fundamental mode is all that can be identified.

It is evident that observing a building's response to wind loading and selecting a section of sinusoidal motion is not the most scientific way of measuring the fundamental frequency, but it can be used successfully at very little cost. If the motion is recorded on a computer (say for approximately a quarter of an hour) and then analysed using an FFT procedure, the frequency of the fundamental modes should be identified clearly, and in this case the frequencies of higher order modes may become apparent (see Figure 4.5). This form of analysis is dealt with later.

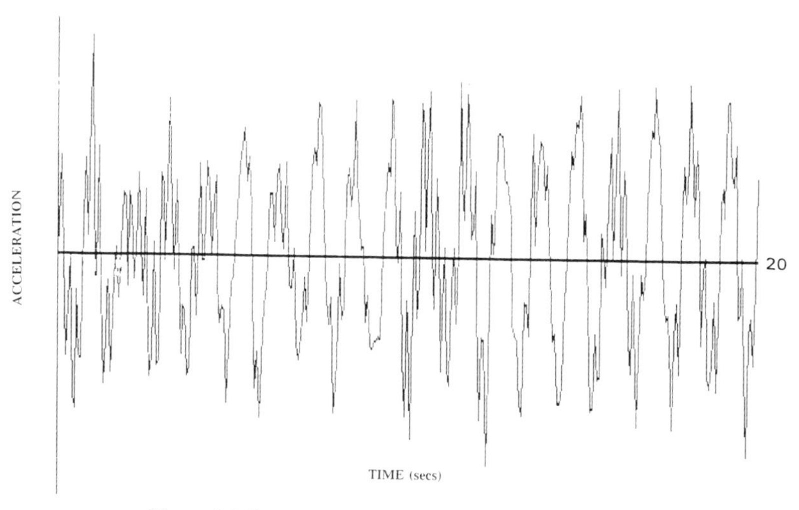

Figure 4.4 Response of a tall building to wind excitation.

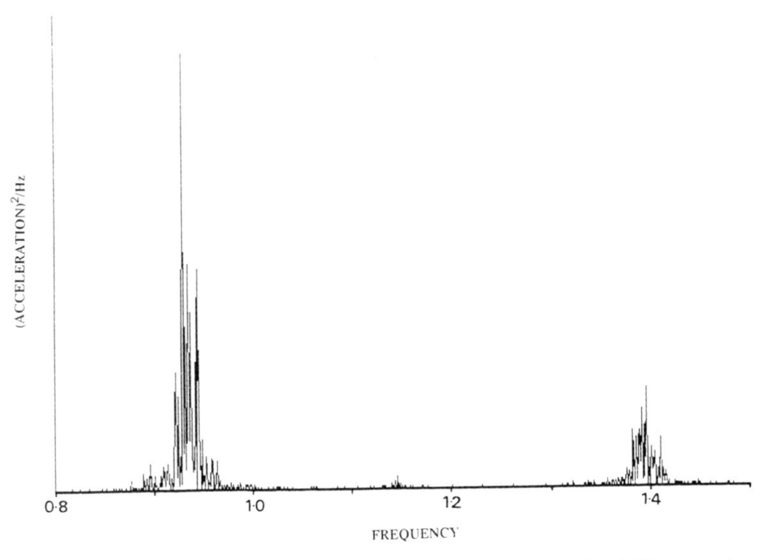

Figure 4.5 Autospectrum of response of a tall building to wind excitation (1024-second sample digitized at 16 Hz).

4.3.4 Damping from decays

Whereas measuring the natural frequency of a fundamental mode of a structure can be relatively simple, estimating the damping is far more difficult. Initially consider a simple problem, that of measuring the damping of a simple steel column. If the top of the steel column is given a small displacement, released and the motion recorded (say using a geophone), then a decay of oscillation similar to that in Figure 4.6 will be obtained. The frequency can be estimated as in the previous section, and the damping can be estimated as the shape of the decay is defined mathematically. The simplest method is to measure the amplitude of N successive peaks in the decay (say a_0 to a_N) and use the formula given in the figure to estimate the damping.

This example shows how damping can be estimated from an experiment where the structure is relatively simple with very low damping. Most structures tend to exhibit behaviour which is not so simple. The response of a floor subjected to an impact may give a decay of similar form to the decay from the column, in which case it is possible to estimate damping. More often, a more complex decay will be obtained which is a combination of several modes of vibration. Given computing facilities it may be possible to filter out the higher modes to leave a decay similar to that in Figure 4.6 and damping can again be estimated, although experience suggests that this is likely to produce slight overestimates. Sometimes even filtering out the higher frequencies will not yield a reasonable decay, and if the decay does not look like a gradually

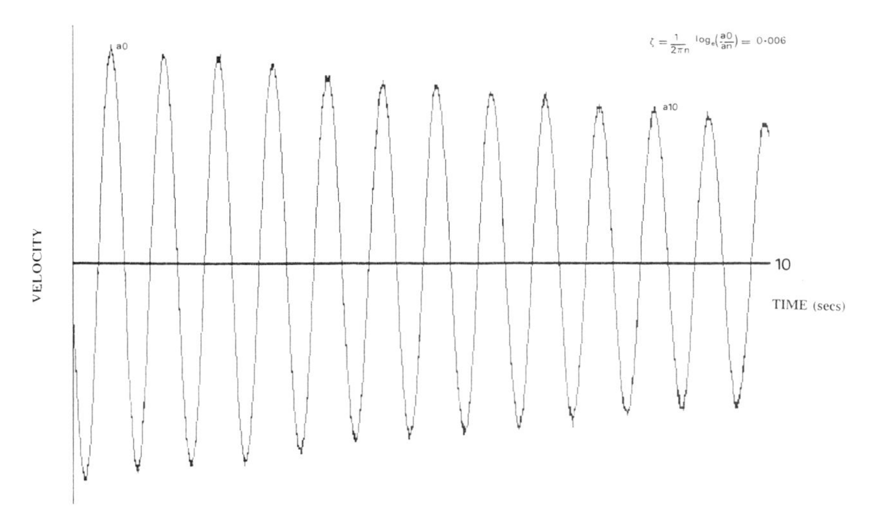

Figure 4.6 Decay of vibration of a steel column.

decaying sinusoid it should not be used to estimate damping. In one building, impact tests on a large number of floors yielded decays which were acceptable for damping measurements in approximately 50% of the cases. Similar tests on walls produced only a few decays suitable for estimating damping. Consequently more refined techniques are required in the majority of cases.

4.3.5 Damping from spectra

Figure 4.7 shows a spectrum of the response of a single degree of freedom system to white noise, i.e. constant force at all frequencies. This curve is defined by three parameters, stiffness, natural frequency and damping. If the curve can be derived experimentally it can be used to evaluate the three parameters. The natural frequency is approximately the frequency of the maximum response. The stiffness can be found knowing the displacement and force, i.e. a knowledge of the input force is required to derive stiffness but frequency and damping can be obtained from the shape of the curve. The damping defines the shape of the curve and amplification at resonance. The usual way of estimating damping is termed the half-power bandwidth method and is shown in Figure 4.7. It involves measuring the response at resonance R and determining the frequencies f_1 and f_2 at which the response is $R/\sqrt{2}$. These frequencies are then used in the equation to determine the damping:

$$\zeta = \frac{f_2 - f_1}{f_2 + f_1}$$

More complex algorithms can be used to make use of all the data points in the curve to provide a better estimation of the damping.

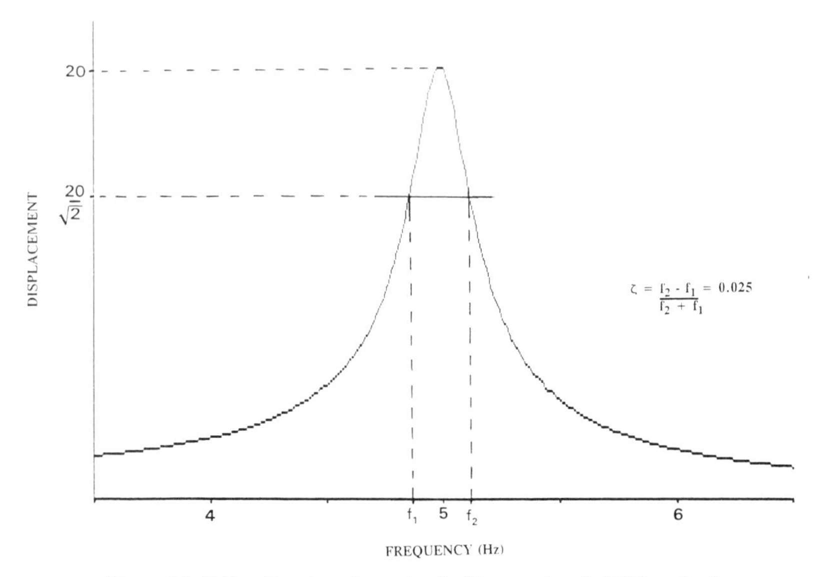

Figure 4.7 Estimating damping using 'half-power bandwidth' method.

There are two principal methods of obtaining such a spectrum experimentally. The first is from a forced vibration test when response measurements are taken at discrete frequencies, and the second is using spectral analysis techniques when the structure is subjected to a constant force over a range of frequencies from an electromagnetic vibration generator or measuring the response of the structure to ambient excitation, e.g. the wind. Whilst the results from forced vibration tests are deterministic and can be expected to be accurate, the spectral analysis is statistically based, requires certain characteristics to be exhibited by the data, e.g. stationariness, and has certain errors which should be considered. Both forced vibration tests and spectral analysis are dealt with later. However, it can be said that spectral analysis from ambient data recorded in buildings is often misused and almost always overestimates damping. Unless there are good reasons to believe otherwise such data should be treated with caution.

4.3.6 Damping from forced vibrations

With forced vibration tests an exciter is used to vibrate a structure. If the exciter is used to vibrate only the fundamental mode of vibration and the vibration is suddenly stopped, a good quality decay of vibration should result. Because the excitation is only in the fundamental mode, other modes of vibration which would complicate the decay, as in the impact tests, are not present. This procedure for exciting the fundamental mode of vibration, suddenly stopping the excitation and monitoring the ensuing decay, is perhaps the most reliable method for estimating damping, and unlike most other methods of estimating damping it is relatively free from potential errors.

The author has used this method for measuring damping on a wide range of structures and structural elements, and Figure 4.8 shows one example of a decay recorded for the fundamental mode of a long-span lightweight floor. Problems occur with this method when there are other forces (maybe wind on a tall building) acting on the structure, or if the exciter does not stop suddenly, in which case it is important to be sure that the decay is that of the structure not the exciter slowing down. The method often cannot be used to get accurate measurements of higher frequency modes because lower frequency modes are usually excited when the exciter is stopped since an exciter with a significant rotating mass cannot be stopped instantaneously.

An important point to realise is that although a number of different methods are available for estimating damping, a lot of them can be easily misused and this almost inevitably leads to an overestimation of damping. This has resulted in many textbooks containing damping data which are much too high, and it will be many years before this problem is rectified. Of course, this does not imply that the methods are theoretically incorrect, but in practice they are often misused. Decays of vibration from fundamental modes are relatively free from potential errors, and when compared with other methods of measuring damping when applied correctly (usually a much more arduous task) yield similar results.

4.3.7 *Stiffness and mode shape from forced vibration testing*

To measure the stiffness and mode shape of a mode of vibration, a comprehensive test set-up is required. This involves the use of one or more

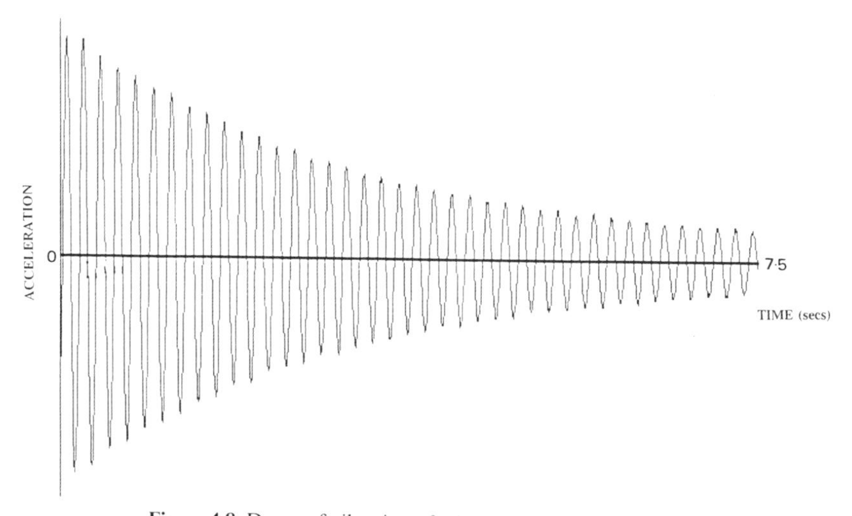

Figure 4.8 Decay of vibration of a long-span lightweight floor.

vibration generators to vibrate the structure in a controlled manner. Various different vibration generators are available which can cover a wide range of frequencies and power output. The vibration generators used by the author are simple in concept and use a pair of contra-rotating masses to provide a unidirectional sinusoidal force. These are described in more detail in the section dealing with instrumentation.

The test scheme used by the author is basically the same whether the test specimen is a wall, a floor or a complete building, albeit different sizes of vibration generator are required. The general test scheme is as follows:

(1) The vibration generator is rigidly attached to the structure at a position where the maximum modal force is likely to be exerted, i.e. at the top of a building or in the centre of a floor.
(2) A transducer is placed near the vibration generator to monitor motion in the desired direction. The output from the transducer is amplified (if necessary) and filtered before being displayed on an oscilloscope and fed into a computer.
(3) The generator is adjusted to produce the desired force level.
(4) The frequency of excitation is incremented through a selected range and the response noted at each stage.
(5) The plot of frequency versus response is normalised by converting the measured motion to equivalent displacements and then dividing the displacement by the applied force. The plot of normalised response against excitation frequency is herein termed the response spectrum.
(6) The natural frequencies are identified by peaks in the spectrum. The actual values of the natural frequency, damping and stiffness for each mode are determined by fitting a multi degree of freedom model to the response spectrum.
(7) A further measurement of damping for each mode is obtained by suddenly stopping the excitation at the natural frequency and analysing the resulting decay of oscillation.
(8) The mode shape for each mode is obtained when the structure is subjected to steady-state excitation at its natural frequency, by moving a second transducer to various positions on the structure and comparing the response with that measured using the reference transducer.
(9) The whole procedure is repeated to monitor modes in other directions and also to examine the effects of varying the force generated.

The response spectrum obtained for the tests on a floor is shown in Figure 4.9. The crosses on the figure represent the measured values and the continuous line represents the best-fit two degree of freedom curve. The values of frequency, damping and flexibility are the parameters which define this curve and hence provide the best estimate of the equivalent parameters defining the modes of the structure. The analysis of the decay provides an independent measurement of the damping parameter, although it is often

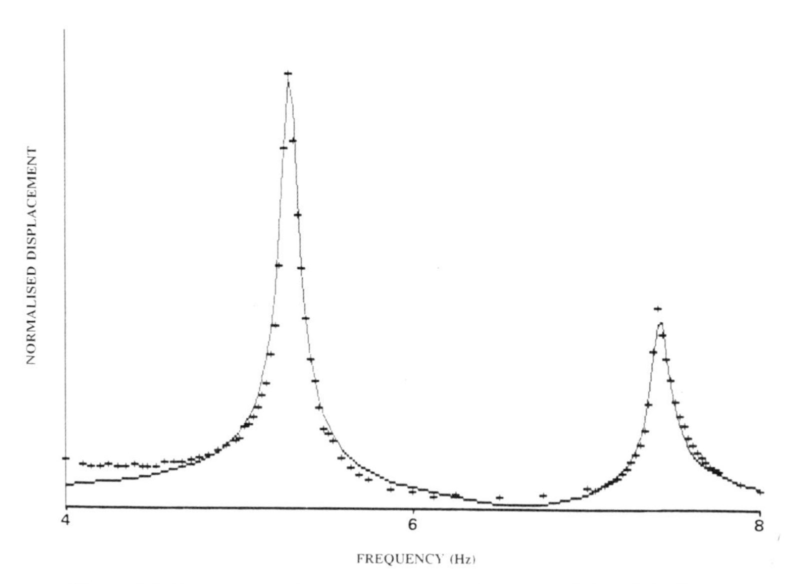

Figure 4.9 Response spectrum measured on a long-span lightweight floor.

only useful for the fundamental mode. For the example presented the damping from the decay measurements was 0.86% (Figure 4.8) and the damping from the curve fitting was 0.87%, which gives an indication of the accuracy of the measurements.

4.3.8 Spectral analysis

Instead of introducing an artificial force to excite the structure it is possible to monitor the response of the structure to naturally occurring forces and identify certain structural parameters. This involves recording the motion of the structure and analysing the recorded response on a computer using spectral analysis techniques. Most spectral analysis techniques use statistical methods and require the data to form a stationary sample. That is, their key statistical properties do not vary over the record. If the record is stationary there are two types of error associated with the estimation of the spectral ordinates, and these are termed variance and bias errors. The limitations of two types of errors constrain the lengths of data which should be used. For a tall building subject to wind loading this may mean that several hours of data have to be recorded to limit the errors to, say, 10%, yet for these data to be stationary the wind speed and direction must have been constant over the whole record. Consequently it can be appreciated that sufficient lengths of stationary data are rare, and in fact in most tests on buildings stationary data are not encountered.

Consequently, some powerful statistical methods exist which require stationary data, but stationary data are unusual when testing buildings. Hence a situation exists which is open to abuse, especially as the analysis instrument will produce an answer whatever the status of the input data. Naturally there are ways of overcoming some of the difficulties and analysing non-stationary data, but these are very much in the realm of the specialist and of limited use in buildings.

Despite these words of warning, there is one very important parameter which can be identified with confidence from non-stationary data, and that is a natural frequency. This is because the frequencies do not vary significantly with gradually changing excitation although, of course, the amplitude of the response will.

The principal use of spectral analysis is to take a record of structural response recorded over a chosen period of time and transform this record to produce a spectrum of response against frequency. The most useful tool for performing this transform is the fast Fourier transform (FFT). As the FFT produces real and imaginary terms, it is useful to manipulate the data further to present them in a more easily understandable format. One such method involves calculating the autospectral density function (autospectrum), which is also often called the power spectral density (PSD). This in effect is a summation of the squares of the real and imaginary terms and compensates for terms which are attributed a negative frequency in the transform. Two autospectra have already been presented, one for an impact and one for a short length of data recorded on a building. In both cases the fundamental mode is clearly shown, and these are typical examples of the data which can be obtained for buildings. However, for these two examples it would be unreasonable to attribute any meaning to the amplitude of the response or to the exact shape of the curve.

There are a couple of basic relationships which are worth mentioning regarding this type of transform. The spectrum is actually evaluated at a number of discrete points; for a full autospectrum this is half the number of points in the original digitised response–time history. Secondly, the frequency range of the spectrum is from zero to half the frequency used to digitise the original recording. Combining the two relationships determines the frequency steps between the spectral lines. Hence for a given sampling rate, the longer the record, the closer the spectral lines or the better the resolution.

One point which should be considered is the possibility of aliasing when digitising a signal. If a sinusoid of frequency higher than $1/2t$ Hz is sampled at a rate of $1/t$ samples per second, then the sinusoid will appear as a lower frequency, but this will not be distinguishable from real data at this frequency. The frequency of $1/2t$ is called the folding frequency. To prevent this occurring the input signal should be filtered before digitisation. Low-pass filtering of the input signal at 1/4 of the sampling frequency is recommended.

Although this section has dealt primarily with frequency measurements, spectral analysis can have many uses when stationary data are available for analysis, and spectrum analysers have many powerful facilities for use by the specialist. These are used extensively for the modal testing of cars and aircraft and are often combined with complex finite element analysis. However, as yet they have little application within buildings.

In one experiment by Littler and Ellis (1989), response data were recorded on a building subjected to wind excitation which enabled spectral analysis techniques to be used to provide more data than just f equencies. On this occasion a vast number of data were recorded (over 2200 hours), the data being split into 1024-second blocks and transformed to produce spectra. Each block was labelled with the wind speed and direction, and when enough blocks were available blocks within selected wind speed and direction limits were averaged. This is attempting to force the data to be stationary, and if, say, 50 blocks are averaged a spectrum like that in Figure 4.10 can be obtained. This can be compared with Figure 4.5, which is a spectrum obtained from just one block, i.e. 1024 seconds, with similar wind characteristics. This shows what effort is required to obtain statistically accurate data, but for specialist research work the results can justify the investment and produce a great deal of new information.

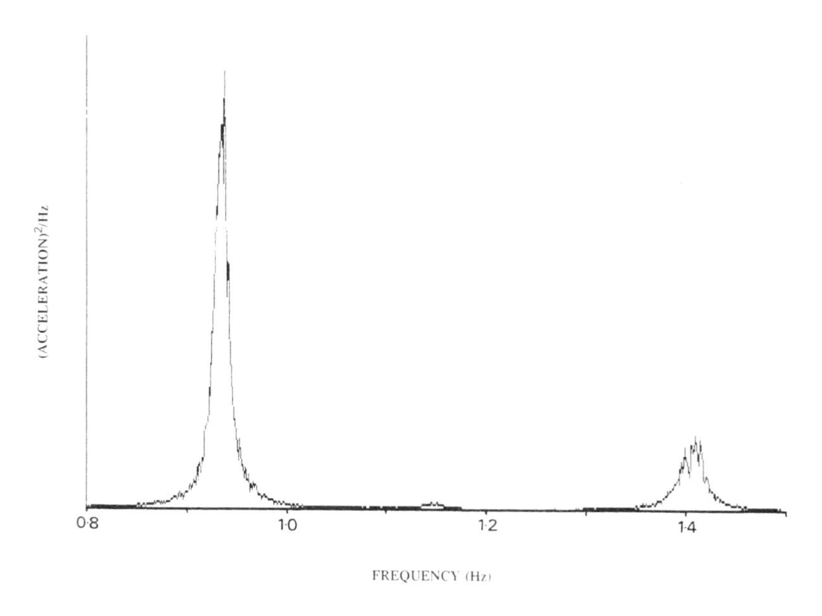

Figure 4.10 Autospectrum of response of a tall building to wind excitation (50 × 1024-second blocks—for comparison see Figure 4.5).

4.4 Instrumentation

The monitoring system usually consists of one or more transducers, some signal conditioning and a recording/analysing instrument. Whatever the individual instruments used it is important to be sure of the calibration of the system. The whole measurement chain can be calibrated on site or the individual instruments can be calibrated in the laboratory using a known source prior to use.

Many different types of transducer can be used, although they are generally used for acceleration, velocity or displacement measurements. As a rule it is best to measure the required parameter directly rather than measure it indirectly and then correct it mathematically, e.g. measure velocity directly rather than integrate acceleration. The parameters which are usually measured are acceleration or velocity. Accelerometers are the transducers used to measure acceleration. They are available to cover a wide range of frequencies and acceleration levels, but top-quality accelerometers may be both expensive and delicate. Geophones are the transducers often used for measuring velocity. They are commonly used for taking geophysical recordings and are often cheap and robust. They will not be found to cover such wide specification ranges as accelerometers, but will be acceptable for many types of measurement within buildings. Displacement measurements usually require part of the transducer to be attached to a static reference position and may prove difficult to use in buildings. Two types of transducer are used for displacement measurements: first the LVDT (linear variable displacement transformer), which has a moving section attached to the structure and the main body attached to the reference position, and second the proximity transducer, which is a non-contact device. Other transducers which can be used are lasers and strain gauges. If a contact transducer is used it should be positively attached to the structure to be monitored, which in many cases requires it to be attached directly to the structure or to a very stiff fixing bracket.

Whatever transducer is used it will require to have the appropriate conditioning instrumentation. Accelerometers and LVDTs may need a power supply, whereas geophones usually provide a signal which can be fed directly into an oscilloscope. Depending on the amplitude of the output signal it may be necessary to amplify the signal to accommodate the range required for the recording device, and it may be necessary to filter the signal to remove unwanted frequencies. If the signal is to be recorded digitally then it will be necessary to provide some filters to prevent aliasing.

The signal should then be ready for recording or analysis. It is usually advantageous to display the signal on an oscilloscope, and in many cases this is all that is required. Nearly all oscilloscopes will have specifications far in excess of that required for monitoring vibrations in buildings, but a dual-channel storage oscilloscope is ideal. Not too long ago it was common to record the signals on y–t plotters to obtain a 'hard' copy or on analogue tape recorders, but the progressive development of computers now provides a powerful and not too expensive alternative. To record directly onto a computer requires analogue to digital (A/D) cards to be fitted into the computer and the appropriate software to display and analyse the signal. As this is a very competitive area for manufacturers, good-quality systems are available at reasonable cost. Some expensive specialist analysers are also used, but they usually have many facilities which will not be required for monitoring buildings.

The final instrument required to enable detailed tests on a structure to be undertaken is a vibration generator. Many electromagnetic and hydraulic vibration generators are available, but for the tests described herein the author has used a range of mechanical vibration generators. The mechanical generators are based on a very simple principle, that of rotating an eccentric mass at precise frequency. The rotating mass produces a tangential force equal to $mr\omega^2$ where m is the mass, r is the distance between the centre of mass and centre of rotation and ω is the angular frequency. The generators used by the author actually use two contra-rotating masses so that the forces from the two masses are additive in one direction and cancel out in the orthogonal direction, thus producing a sinusoidal unidirectional force. This type of generator could be made in any good workshop. The frequency control is obviously of prime importance, and the wide range of stepper motors which are commercially available provides adequate power and control for all but the biggest generators.

Naturally, the bigger the structure to be tested, the larger the force required and the bigger the generator. The force and frequency ranges of the three vibration generators used by the author are:

Large system	0.5–20 Hz	max peak–peak force of 8.2 tonne
Intermediate system	0.2–20 Hz	max peak–peak force of 0.2 tonne
Small system	5.0–120 Hz	max peak–peak force approx 0.11 tonne
		min peak–peak force $0.0026 \times f^2$ N

The large system consists of four individual vibration generators, one of which is shown in Figure 4.11. The system has been used to test a range of large structures, the largest of which was a 200 m tall arch dam. This is the system used to test tall buildings and can only be used to provide horizontal excitation. Each vibration generator requires four people to lift and manoeuvre it into position. The intermediate system can be used vertically and is used mainly for long-span floors, although it is quite adequate for testing a building. The generator can be split into two sections, the largest of which requires two people to carry it easily. The small system is used for testing walls and floors in buildings and is usually used with the lower force setting. The actual generator is very small and can easily be carried in one

Figure 4.11 One of the four Building Research Establishment large vibration generators.

hand, although the associated stepper motor and control are heavier but can still be managed by one person.

4.5 Use of information

4.5.1 Scope

So far this chapter has dealt with how to measure various structural parameters and how accurate the measurements are likely to be. This section deals with how the information can be used.

4.5.2 Frequencies

One of the most important aspects of testing an existing building is to provide information to verify that it is behaving in a similar manner to that assumed in its design. A measurement of the fundamental frequency of a structure or structural element can be compared with a calculation of the same parameter to provide a check on the accuracy of the calculations. As the frequency can be calculated from the simple relationship between stiffness and mass, it can be appreciated that this is a verification of the model used to design the structure. If the designer cannot work out the stiffness and mass of the structure, then he will not know how the loads are carried by the structure or what stresses are encountered or whether the stresses are acceptable.

It is not always possible to calculate the fundamental frequency of a structure accurately, and providing feedback to check calculations is useful for two reasons: firstly, in assessing accuracy of the assumptions and calculations used in the design of that particular structure and secondly in learning how accurate mathematical models are likely to be in general. This second point will enable weakness in design assumptions to be identified, it will provide data to enable research to concentrate on real problems and so will lead to a better understanding of structural behaviour.

One problem that the author has encountered on many occasions is that it is often assumed that calculations are far more accurate than they really are. If, for example, the reader was asked to estimate the accuracy with which a typical finite element model could be used to calculate the fundamental frequency of a tall building, would he estimate within 1%, 5%, 20% or worse? Remember that buildings are actually very complicated structures. Ellis (1980) surveyed publications in which both computer-based calculations and measurements of the fundamental frequencies of tall buildings were given. From this survey a sample of data was selected for rectangular plan buildings where calculations were not influenced by measurements. Whilst it was thought that this might be a somewhat biased sample (perhaps only the best results get published?), it turned out that the simple empirical relationships were more accurate than computer-based calculations. With empirical calculations errors of more than $\pm 50\%$ are not uncommon, so it can be seen that predictions of fundamental frequencies may not be too accurate.

With some structures, the fundamental frequency is required to check a serviceability limit. These are structures in which vibrations from human actions, e.g. walking or dancing, may be sufficient to annoy other users of the structure. These are typically lightweight floors or foot bridges, and a simple rule that is often given is that the fundamental frequency should be above 5 Hz. This figure reflects the fact that human excitation cannot generate a significant energy input at frequencies above 5 Hz. An experimental check of the fundamental frequency has obvious merits, especially if the frequency limit was a significant factor in the design of the structure.

4.5.3 Damping

Damping may appear to be of little use on its own. Indeed, if the literature reflected typical (and accurate) damping values for a wide range of structures then measuring damping by itself would have little meaning. Damping is actually a parameter which is important in resonant vibrations. If a case of resonance occurs then damping is a very important parameter in that it controls the amplitude of the response and the time taken (or number of cycles) to reach that amplitude.

For cases where resonance may occur, methods of calculation are often given which require a damping parameter. For example, methods to determine the response of floors to vibration require a damping parameter (for the fundamental mode). As this parameter is usually inversely proportional to the calculated response it is important. For a number of years, a number of Canadian papers provided the state of the art in this area. From these publications damping for bare composite floors would be expected to be around 2%, increasing to 5% for a finished floor plus furniture. Recently a British design guide has appeared which suggests values of damping of 1.5% and 3% for situations similar to those defined by the Canadians. However, when Osborne and Ellis (1990) tested one particular floor, damping values below 1% critical were measured for the bare floor and also for the floor with false flooring and services added. Naturally this is just one measurement, which might be atypical, but also it might be that previous measurements have overestimated damping. However, it has a major effect on the results of the calculations and, in the author's opinion, if it is worth doing the calculations then it should be worth checking certain key parameters in the prototype structure, especially when the basic data have been questioned.

4.5.4 Mode shapes

Mode shapes are usually measured in forced vibration tests, although they can be obtained from ambient response measurements. The mode shape can be useful along with the measured frequency for checking calculations, because accurate calculations should predict the correct frequency and corresponding mode shape. Alternatively if the frequency prediction is incorrect, then the mismatch in mode shape might indicate where the assumed stiffness distribution is incorrect. One interesting example occurred when Brownjohn *et al.* (1987) compared calculations of frequencies and mode shapes of a suspension bridge with measurements. In this case, two models using different support conditions predicted similar fundamental frequencies but yielded different mode shapes. The mode shape was therefore used to identify which model was correct.

One area where difficulties often arise is in modelling the restraint conditions for a particular structure. For example, it is often assumed that the base of a tall building is rigid, however the measured mode shape might indicate movement at the base of the building, indicating that soil–structure interaction is significant. Alternatively, the problem can be inverted, in that measured mode shapes (along with stiffness and frequency) can be used to measure the effect of the restraints. For tall buildings these types of measurement have been used by Ellis (1986) to evaluate the effective soil stiffnesses and thereby check theoretical predictors of soil–structure interaction.

4.5.5 Stiffness

When a forced vibration test is conducted and the frequency, damping, mode shape and stiffness of each relevant mode of vibration have been measured, then the system is well defined. These are the parameters required to predict the dynamic (or static) behaviour of a structure within the elastic region. For example, these parameters are required in conjunction with the actual wind load to predict how a building will respond to wind loading, and an accurate knowledge of these parameters is obviously a good basis for the calculation.

To demonstrate this point it is worthwhile examining one example in detail. Consider how a wall of a building would respond to an internal gas explosion. A gas explosion is long in comparison with the natural period of a typical wall, therefore the pressure distribution over the wall will be effectively uniform and the response will be primarily in the fundamental mode (albeit effectively static response not at the fundamental frequency). Measurements of the characteristics of walls in a number of buildings show that the edge restraints vary significantly and are extremely difficult to predict. At the Building Research Establishment's explosion rig at Cardington, a wall panel was erected and subjected to a number of non-destructive explosions by Ellis and Crowhurst (1989). The panel displacement and the pressure time histories were recorded. To predict how the panel should respond, the panel's characteristics were first measured in a forced vibration test. These characteristics were then used with the measured pressure–time history to predict the panel response. Two different forms of calculation, the Duhamel integral method and a predictor–corrector method, gave identical results, and as can be seen in Figures 4.12 and 4.13 the calculations and measurements are similar.

This correlation between calculation and measurement was obtained for numerous explosions and is a result of defining the actual structural system accurately. When this type of calculation is undertaken it is essential to remember that the measurements from the vibration tests yield modal characteristics, and to perform the calculation it is necessary to calculate the modal force, i.e. pressure × area × mode shape (see Appendix). Also, it must be pointed out that this is elastic behaviour, and plastic and ultimate behaviour are another story.

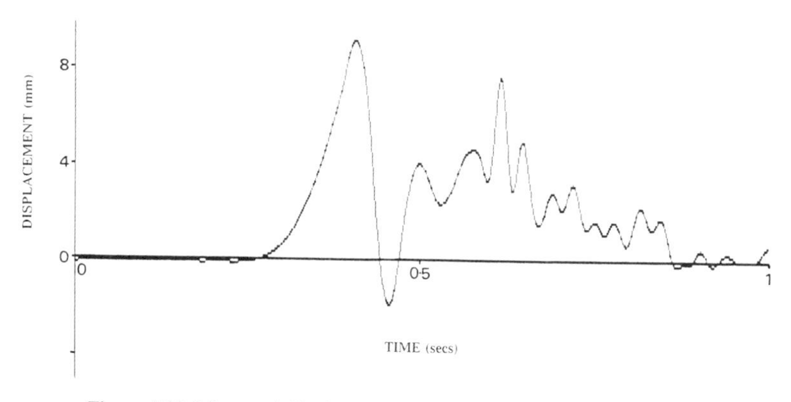

Figure 4.12 Measured displacement of a panel during a gas explosion.

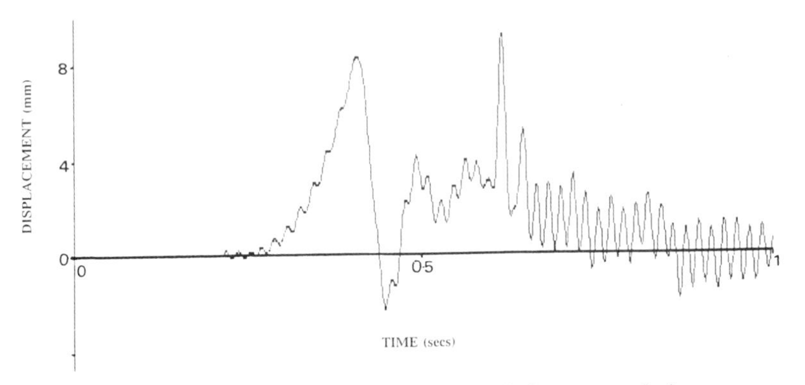

Figure 4.13 Predicted displacement of a panel during a gas explosion.

4.6 Non-linear behaviour

It has been mentioned that structures do exhibit certain non-linear characteristics, and it is worthwhile considering what is meant by a linear material or a linear system. A material is considered to be linear if its Young's modulus is constant, i.e. if stress is directly proportional to strain. However, if the material is damped the plot of stress against strain would exhibit a hysteresis loop, or part thereof, so that no real material is truly linear. The behaviour of a system can often be described by a differential equation, and the system is called linear if the influence coefficients related to the differential terms are constant. For the behaviour of buildings to be termed linear requires both the damping and natural frequency of each mode to be constant.

The damping values and natural frequencies which are observed in practice do vary with the amplitude of motion, and to demonstrate these non-linearities it is worth while examining the results obtained from tests on one building. The results are shown in Table 4.1. This was a building where soil–structure interaction was significant, but the characteristics are similar for buildings where soil–structure interaction is negligible. There are two basic relationships to be seen in the table; firstly the resonance frequency decreases with increasing amplitude of motion, and secondly the damping increases as the amplitude increases. It can be seen that over the range of the tests (which correspond to motions produced by winds varying from light to hurricane) the change in natural frequency in the north–south (NS) mode is relatively small (3%), whereas the change in damping is much larger (30%). It is also interesting to note that if the amplitude (or corresponding acceleration) was plotted on a logarithmic scale, then the two relationships would be approximately linear, i.e. frequency vs log amplitude and damping vs log amplitude. This points to there being a common mechanism producing the non-linearities.

Table 4.1 Results of non-linear tests at Dunstan Mill.

Mode	Frequency (Hz)	Peak–peak force (kN)	Peak–peak response ($\times 10^{-3}$ g)	Damping (% critical)
NS1	1.45	18.0	23.37	2.60
	1.46	13.7	17.94	2.43
	1.47	9.3	12.36	2.42
	1.48	4.7	6.59	2.29
	1.48	2.9	4.17	2.18
	1.49	2.0	2.96	2.11
	1.49	1.1	1.60	2.02
EW1	2.21	41.8	33.94	3.24
	2.22	31.7	26.18	3.16
	2.23	21.3	17.94	3.00
	2.25	10.8	9.45	2.94
	2.26	6.7	5.91	2.85
	2.27	4.6	4.12	2.69
	2.28	2.5	2.42	2.57
		(KN.m)	(rad/s $\times 10^{-3}$)	
$\theta 1$	3.90	272.8	19.02	3.07
	3.92	176.4	13.16	2.79
	3.94	96.4	8.02	2.44

Although two approximately linear relationships are presented it is not wise to extrapolate significantly using the relationships. The building was being tested in the non-linear elastic region; however, if the amplitude of vibration is sufficiently large, it is probable that the stiffness deflection characteristics will change from non-linear elastic to elastoplastic, i.e. similar to many materials when a yield point or maximum damping capacity is reached, and hence produce irreversible changes, i.e. damage.

For most cases the non-linearities can be ignored, although it is useful to be aware that they exist. Structural characteristics have also been observed to vary with temperature and even tide height for some buildings adjacent to tidal rivers.

4.7 Integrity monitoring

The use of non-destructive tests (NDT) to identify the integrity or condition of a structure or structural element has been the subject of considerable interest internationally. The ability to test a structure and identify key parameters which can be related to overall behaviour, potential damage or other problems has obvious attractions; and although this may be conceptually quite simple, many practical difficulties arise in testing real structures.

First consider the problems involved with assessing damage from a NDT.

NDTs using vibration measurements can provide good measurements of the elastic behaviour of a structure, i.e. measurement of the characteristics of the modes of vibration. Damage to structures sometimes results in a reduction in stiffness and hence a change in frequency, and it has often been suggested that monitoring frequency will identify when damage occurs. However, frequencies also vary with temperature and other factors, and it is very difficult to separate effects due to damage from other effects, especially when the effects cannot be predicted accurately. Also damage may not necessarily lead to measurable changes in stiffness, hence it is not usually practical to try to assess damage from this type of measurement.

However, the idea of monitoring the structure in order to provide feedback about its behaviour has many merits. The gain from the monitoring is achieved when mathematical models are compared with measurements, thus providing more confidence in calculations. Knowledge of the observed behaviour may also be useful for serviceability assessments and can help to improve estimates of fatigue damage in certain instances.

Appendix: Basic background theory

Coordinate transformation and calculation of modal force

Some quantities which are measured are modal quantities (resonance frequencies and damping ratios), whereas others (accelerations, displacement and pressures) are not. For consistency in calculations the non-modal quantities should be converted to modal quantities. The modal approach is used because it offers the advantage that each mode can be considered independently as a single degree of freedom system, unless two modes are so close together that they appear coupled.

The matrix equation which describes the motion of a linear viscoelastic structure using Cartesian coordinates is:

$$M\ddot{X}+C\dot{X}+KX = P(t) \tag{A1}$$

where M is the structural mass matrix, C is the structural damping matrix, K is the structural stiffness matrix, $P(t)$ is the structural force vector and X is the vector of generalised displacement.

This equation is found in every textbook dealing with structural dynamics and provides a good description of the behaviour of real structures in the elastic range. This equation represents a series of coupled equations. If the damping and force matrices are set to zero, so that the undamped free vibration of the system is examined, it is possible to uncouple the equations by finding the eigenvalues of $M^{-1}K$. This actually yields the eigenvalues which are frequencies of the various modes of vibration and the eigenvectors which are the corresponding mode shapes.

Eqn (A1) can be transformed from Cartesian to modal coordinates using:

$$X = \phi x$$

where ϕ is the generalised mode shape matrix and x is a vector of the modal displacements. If the equation is then premultiplied by ϕ_n^T, the transpose of the mode shape vector for the nth mode, the following equation is obtained:

$$\phi_n^T M \phi \ddot{x} + \phi_n^T C \phi \dot{x} + \phi_n^T K \phi x = \phi_n^T P(t) \qquad \text{(A2)}$$

The standard orthogonality conditions which can be applied to the above equation state that:

$$\phi_n^T K \phi_q = \phi_n^T M \phi_q = 0 \text{ for } n \neq q$$

If it is assumed that the modes are orthogonal with respect to C, which is reasonable because distinct values of modal damping are observed, then:

$$\phi_n^T C \phi_q = 0 \text{ for } n \neq q$$

The scalar values $\phi_n^T M \phi_n$, $\phi_n^T C \phi_n$ and $\phi_n^T K \phi_n$ can be expressed as m_n, c_n and k_n and the force $\phi_n^T P(t)$ as $p_n(t)$ so that eqn (A2) can be represented by a series of independent equations of type

$$m_n \ddot{x}_n + c_n \dot{x}_n + k_n x_n = p_n(t) \qquad \text{(A3)}$$

where x_n is the modal displacement in the nth mode.

It should be noted that ϕ is a dimensionless vector describing the mode shape and it can be normalised to any chosen value without affecting the validity of the above equations. Each of the n equations represents a single degree of freedom system expressed in modal coordinates, and the usual approximation holds for small damping:

$$\omega \simeq \sqrt{\frac{k}{m}}$$

and at resonance if $p_n(t)$ is sinusoidal at frequency ω with maximum value p_n:

$$k_n = \frac{p_n}{2\zeta x_n}$$

where

$$\zeta_n = \frac{c_n}{2\omega_n m_n}.$$

References

Brownjohn, J.M.W. *et al.* (1987) Ambient vibration measurements of the Humber suspension bridge and comparison with calculated characteristics. *Proc. Inst. Civ. Engrs.* pt 2, Sept.

Ellis, B.R. (1980) An assessment of the accuracy of predicting the fundamental natural frequencies of buildings and the implications concerning dynamic analysis of structures. *Proc. Inst. Civ. Engrs.* pt 2, June.

Ellis, B.R. (1986) Dynamic soil–structure interaction in tall buildings. *Proc. Inst. Civ. Engrs.* pt 2, June.

Ellis, B.R. and Crowhurst, D. (1989) On the elastic response of panels to gas explosions. *Proceedings of the conference on Structures under Shock and Impact, Cambridge, Massachusetts.* Elsevier, Amsterdam.

Littler, J.D. and Ellis, B.R. (1989) Interim findings from full-scale measurements at Hume Point. *6th US Wind Conference*, published in *Journal of Wind Engineering and Industrial Aerodynamics*, **36**, 1181–1190.

Osborne, K.P. and Ellis, B.R. (1990) Vibration design and testing of a long span lightweight floor. *Structural Engineer*, 15 May, **68**(10), 181–186.

5 Automatic and autonomous monitoring

A. KENCHINGTON

5.1 Introduction

5.1.1 Purpose

There are many instances in which traditional techniques for the monitoring of building structures can be unsatisfactory or unsuccessful. Traditional techniques are necessarily labour intensive and are, as a result of economic constraints, frequently incapable of producing a readily interpreted body of data. In difficult cases, building owners, operators and managers are not likely to be happy to meet the costs of engaging skilled personnel for extended periods even when they can accept the intrusion that the use of traditional techniques necessarily involves. When the phenomenon of interest requires close attention over a (moderately) extended period or when monitoring must be done in a discreet fashion, automatic and autonomous monitoring of building structures comes into its own.

These factors were involved at St Paul's Cathedral in London where automated monitoring systems were installed to monitor cracks in the South Transept and movements of the bell towers. Traditional techniques were producing inconclusive results for deciding what mechanism was responsible for further movement in a network of cracks and in construction joints. The cost of monitoring the situation more closely using manual techniques was prohibitive and the installation of necessary equipment was also going to be very intrusive. Therefore a custom-built monitoring system was installed to rapidly and discreetly gather data which then enabled those responsible to decide that no special action was required in the short term.

In other cases the phenomena which are of interest will be transient in nature. In these circumstances it may be impractical to have personnel permanently in place until such time as the event of interest occurs. Automated, autonomous monitoring can readily distinguish transient and intermittent from continuous phenomena and will guide the adviser towards the appropriate area of enquiry. As an example, evidence of distress in flats above a shopping parade was eventually found to be caused by delivery vehicles turning and mounting a pavement over projecting raft foundations. In this instance, a single bollard prevented further damage whereas initially a serious foundation failure had been suspected.

Monitoring nuisance phenomena is another area where automatic and autonomous monitoring has particular value. Noise and vibration effects are notoriously difficult to quantify and observe. Suitable monitoring equipment can readily distinguish fact from opinion.

5.1.2 Advantages and disadvantages

One of the greatest advantages of automatic and autonomous monitoring systems which are based on currently-available and emerging data-acquisition technologies is the opportunity to eliminate human error from the measurement system. It is now readily practicable to gather data and handle it wholly automatically right through to the point at which it must be interpreted. It is even possible to go one step further and to carry out certain styles of analysis automatically in order to highlight or elucidate subtle trends within a large and complex body of data which may otherwise be missed. The future use of expert systems and artificial intelligence will advance and improve data interpretation methods.

While the replacement of the human factor with a computer system solves many problems, the system itself is nonetheless a human artefact and unless designed and used properly can easily be misused. The design of automatic monitoring systems requires a detailed understanding of the whole of the measurement process and the context in which it is being carried out. It is the purpose of this chapter to give an insight into the options available and the factors affecting a system's performance so that the reader can ensure that a system is provided which will perform as required and produce data which is capable of interpretation in context.

Reading results, recording observations and results in a notebook, and manipulating data manually are all fraught with the risk of human error. A long-term monitoring task carried out using traditional surveying techniques (precise levelling in open air) adequately demonstrated that the quality of the data obtained was highly dependent on factors such as the weather, the motivation and continuity of the personnel involved, and the quality and closeness of supervision. Once installed, and provided appropriate measures have been taken to ensure continuing performance regardless of the prevailing conditions, a properly designed automatic monitoring system will provide an abundant flow of good quality data.

Examining the option to use automatic and autonomous monitoring systems as against traditional techniques highlights the issue of the relatively high first cost of such systems. Automated monitoring systems have to be designed, procured, installed, tested and commissioned before a single piece of data is obtained. Once installed, however, the cost of obtaining a large quantity of data from an automatic system can be very low indeed. Often the problem becomes a question of determining how little data will be sufficient.

In contrast, traditional manual techniques consume resources in proportion to the amount of data obtained and raise the problem of judging how little data 'we can get away with'. The superior data-gathering capabilities of automatic systems enables us to make an informed decision. A traditional manual approach may only permit professional judgement to be made, with attendant indemnity considerations.

An automated system should not be considered as a direct replacement for more traditional techniques. An automatic system will only be of use if its purpose is accurately defined, its scope extends to include a sufficient number of measurement points and parameters, and it provides appropriate accuracy; factors which, in combination, largely determine the first cost of the system. If the design of the system is mis-directed or the scope or accuracy is too limited, then the value of the data produced by the system will almost certainly be greatly reduced. The decision to use automated techniques must therefore be made in the context of sufficient information from traditional sources to enable the purpose of the system to be defined in order that its scope, character and cost can be determined.

Traditional techniques can be inherently more flexible. When manual techniques are employed a programme of monitoring can be redirected more readily as the monitoring progresses and results are obtained. The focus of interest can shift as the behaviour of the structure is revealed in detail. Information which is obtained automatically will provide a quantity and quality of data rarely achieved using traditional techniques. The use of automated systems relieves the burden of making repeated measurements, thus releasing resources for assimilation, appraisal, and analysis (and, when appropriate, for further investigation elsewhere). Automated and traditional techniques are essentially complementary. Now that modular (largely digital) systems are available which are much more readily rearranged and reconfigured than their analogue predecessors, the scope of an automated system can be extended more readily as the nature of the monitoring task is revealed. It is, therefore, practical to think in terms of the two approaches working hand in hand.

The amount and quality of the data which can be obtained from a well designed and constructed automated monitoring system will frequently reveal facets of a building's behaviour which may be otherwise hidden. Automated monitoring systems are normally capable of detecting the response of a building to daily and seasonal temperature changes. These 'signals' may not be of concern in their own right, but they are part of the spectrum within which other movements take place. Using traditional manual techniques these signals may be regarded as only 'background noise' and therefore may conceal what is actually going on. With automated systems capable of recording data at close intervals, it is possible to provide a sound basis for distinguishing cyclical from progressive phenomena, and thus reveal the causes of movements with greater precision.

Movement joints to allow for thermal expansion and contraction of the fabric of St Paul's Cathedral were not provided. The Cathedral has therefore provided itself with appropriate compensating mechanisms by opening and closing joints and fracturing stonework. New patterns of load distribution resulting from ground movements (due to such factors as water extraction in the London Basin and modifications to the way in which rain water is drained from the surrounding streets) have resulted in relative movements within the building, and a need to find new means to cope with diurnal, climatic and seasonal temperature changes. Knowledge of the nature and extent of thermal movements was, and continues to be, vital in arriving at and maintaining a suitably sophisticated understanding of the structure's present behaviour and in making prognoses regarding structural integrity.

5.1.3 Frequency of measurement

An automated monitoring system at Fort Gilkicker (one of the massive forts which forms part of the defences of Portsmouth harbour, England) was installed and commissioned early in 1991 to monitor movements in a system of large cracks. The primary function of the automated system was to establish the character of the movement taking place and to confirm the belief that the movements were not progressive (the cracks were thought to be the result of structural modifications made in the late 19th century). The system rapidly produced evidence of what initially appeared to be movements responding to diurnal temperature changes. Closer examination revealed that the frequency of the cyclical movements was too great and it became apparent that the movement was related to tidal effects. This was confirmed by subsequent work and thereafter it became a relatively simple matter to isolate underlying trends due to climatic and seasonal changes. Furthermore, in the spring of 1991, a rapid change in the relative position of two pieces of stone in the arch forming the opening for one gun emplacement was recorded. This data provided incontrovertible evidence of continuing progressive movement.

A traditional approach would have obtained only a few readings before the movement took place, and there would not have been sufficient data to distinguish fine-scale thermal and tidal movements. The rapid movement of the stonework would probably have occurred between readings, and subsequent readings would simply have recorded a change in the structure. The magnitude of the movement would have been confirmed only after a sufficient number of readings had eliminated the 'noise' in the traditional measurement system, and therefore the precise nature of the movement would not have been determined for some time. The automated system quite clearly identified what had happened almost immediately.

The determination of what is happening or has happened requires a sufficient and suitable body of reliable data. Clearly any system, automated or otherwise, should be capable of resolving the fastest phenomenon of interest, otherwise there is a very real risk that the data obtained will be misinterpreted due to 'aliasing'. Aliasing can occur whenever the frequency of reading is close to or less than the frequency of the parameter being monitored, unless countermeasures are taken.

The situation becomes quite complex when there are other phenomena (which are similar in magnitude) occurring at higher frequency than the phenomenon which is of direct interest. As an example, thermal movement will almost invariably mask the underlying growth in the width of a crack. Thermal shock and related movement resulting from the way the facade of a building is shaded or left exposed to sunlight by surrounding buildings has been seen to alter radically the way a building is perceived to behave after close and precise monitoring.

5.1.4 Range and resolution

Just as the system must resolve with respect to time the phenomena of interest, it must equally well be able to determine the magnitude of the phenomena of interest with appropriate accuracy and discrimination. Differing sensing technologies offer widely different capabilities. The range of measurement between the maximum and minimum inputs which can be measured and the smallest change which can be detected may or may not be linked, depending on the technology involved. In some cases the resolution is unlimited or infinite, though in practice some other part of the system will impose a limit to the resolution. Other technologies are inherently limited. A smooth rise in input will cause a step change in output. Hysteresis, when the measurement obtained depends on whether the input is rising or falling, and (non) repeatability, when the measurement obtained is not always the same for precisely the same input, may also limit the resolution. Some technologies (and all digital ones) offer inherently limited resolution. Some proximity switches used as position sensors are only capable of resolving three states ('near', 'acceptable', and 'far') to take an extreme example. Sensors which incorporate digital filtering techniques to remove unwanted noise from the output signal can only produce a signal which increases and decreases in fixed increments. Furthermore, inaccuracies in the design or manufacture of the electronics may result in missing steps or output states. This will be largely irrelevant if the total number of steps is large in relation to the range. It may well be very important if the step change in the input required to produce a step change in output is large in proportion to the measurement being made.

The system must be accurate enough for the task. Each stage in the process of obtaining an analogue of the input and converting it to either a numerical value or presenting the data graphically, introduces error (or at least the potential for error). Combining data obtained from a number of sensors can compound the error. The system designer must adopt a rigorous approach to determining the potential for error and scale of error which could result in various combinations of circumstances. It is beyond the scope of this discussion to provide the details of how this is best achieved. A sensible practice involves establishing what error can be accepted, reducing this by at least one order of magnitude (to establish the 'error budget') and then determining how this is most effectively apportioned to the different parts of the system. In very many instances it is possible to specify standard devices capable of meeting the requirements. Occasionally this will prove too expensive and it is then necessary to look at the means by which the contribution made by different elements of the system can be reduced using a variety of available techniques. These can include: individual transducer calibration to eliminate non-linearities and offsets in their output; careful control of the environment in which elements of the system are operated to reduce temperature effects; sophisticated techniques for automatically monitoring and adjusting the performance of the data-acquisition equipment itself. Such techniques can considerably enhance the performance of sophisticated and unsophisticated sensing technologies.

A word of caution is necessary. Frequently performance specifications are prescribed which already take account of the potential for error. The data users decide the accuracy required and this is multiplied by a factor to determine the resolution that the system should provide. The factor used may be five, ten or even a hundred, and the data users then quote this figure as the accuracy required. A quotation for a system that will perform to this specification generally exceeds expectations, so users must always be explicit in stating requirements and how they have been determined. The advice of the system designer should be sought regarding what can sensibly be achieved.

Automated monitoring systems of the type considered here, even single channel ones, offer tremendous capabilities. Resolving movement at joints or cracks to hundredths or even thousandths of a millimetre can be achieved quite readily. The data which is obtained is, by virtue of both its quantity and quality, susceptible to very detailed analysis, and as a result fascinating insights into the behaviour of a structure can frequently be gained. Although this may provide good raw data suitable for research purposes, it is generally too detailed for the engineer, architect or surveyor dealing with the client's immediate needs.

The quality of the data which may be obtained provides a temptation to greatly increase the scope and complexity of the system. First cost is in practice a useful restraint. Nevertheless a balance still has to be found as it is equally possible to compromise the whole of a system by limiting its scope or performance too much. The solution lies in the effective management and execution of the processes of determining and defining the requirements, and satisfying the requirements. This chapter intends to provide guidance on this.

5.2 The principal elements of a monitoring system

5.2.1 Overall system

An automated monitoring system consists of a number of sensors and transducers combined with suitable signal conditioning, and connected through an interconnecting medium to data-acquisition and recording equipment. The sensors and transducers convert the parameters of interest into an electrical analogue of that parameter. Signal conditioning provides the means by which the analogue signal is relayed to the monitoring location. Further signal conditioning and processing may be required to convert the analogue signal obtained at the monitoring location into a form acceptable to the data-acquisition equipment. The data-acquisition equipment records and displays inputs from the sensors and transducers in the form of an analogue trace on a chart recorder (or any other analogue device) or as data recorded digitally.

The following sections examine each of these areas in turn and describe some of the options available (and their implications) in order to give the reader an overview of the technology.

5.2.2 Sensors and transducers

The premise for any automated monitoring system is the availability of a suitable sensor or transducer which is capable of providing an analogue of the parameter or phenomenon to be monitored and recorded. Fortunately the range of sensors or transducers available is vast and covers virtually every parameter imaginable. Most have been developed to meet industrial process control requirements, whereas others were developed for medical purposes. The aerospace and defence electronics industries have developed a large number of highly sophisticated sensing technologies, some of which offer extremely high precision—albeit at a very high price. There is therefore a tremendous range of sensors and transducers available to the structure monitoring system designer.

The industrial process control origins of many sensor types bring many benefits and some disadvantages. As the process control industry is a large and mature one, sophisticated equipment is available for relatively modest cost. However, monitoring building structures demands high performance and accuracy, while, in addition, the performance and accuracy have to be obtained with assurance, continuously and usually for extended periods.

In the normal industrial environment sensors can undergo periodic recalibration to ensure continuing accuracy. Alternatively the sensors are employed in systems which incorporate manual or automatic feedback to compensate for 'drift' resulting from wear and tear. At the simplest level, a knob is adjusted when a machine or process operates incorrectly, while more sophisticated systems monitor their own performance. However, automated monitoring systems are subject to much more strenuous demands as they have to be

Figure 5.1 Transducer installation is generally only a problem once. Once installed there is no limit to the number of readings which are obtained.

capable of making measurements over long periods and all of the data collected must be reliable. Quite often it is not possible, practicable, or desirable to demount a transducer once it has been installed in order to carry out calibration checks.

5.2.3 Signal conditioning

Many sensors produce a signal which cannot be reliably transferred over any distance without degradation. Some transducers require quite specific and high quality power supplies or other forms of excitation. In other cases the output from a device is not of a form which is readily interpreted by the monitoring equipment: the range or style of the output from the sensor may be incompatible with the input required by the monitoring equipment. Also, the signal may fluctuate too rapidly to be accurately recorded, and the output from the transducer may consist of a frequency or amplitude modulated carrier whereas the monitoring equipment will only accept a steady voltage or current input. The requirement to make the output from a sensor compatible with means of transfer and the input to the monitoring equipment will generally be met by signal conditioning.

Transducers can be used which incorporate signal conditioning so that, for example, a device can be supplied with DC power and will return a 'high level' signal (a voltage varying between 0 and 10 volts for a current varying between 4 and 20 milliamperes). An alternative method uses modules which can be mounted adjacent to either the sensor or the monitoring equipment.

The same observations about enduring accuracy apply to signal conditioning as to sensors and transducers. When they have not been specifically designed for making measurements in a suitably similar context (i.e. close to the one in which a particular structure monitoring system is to be used), great care must be taken to ensure that the information obtained can be relied upon and will be of an appropriate quality for the duration of the measurement programme.

Signal conditioning will generally provide an analogue signal which can be transferred over considerable distances. The limit will depend on the interconnecting medium used, the precise nature of the monitoring equipment employed, and the context in which it is operated.

Voltage outputs can normally only be transferred over relatively short distances when higher levels of precision are required, and then only if very little current is allowed to flow. Voltages induced by nearby equipment or machinery can seriously degrade the data obtained. AC voltage outputs—such as those which are obtained from some styles of position sensing transducers—are less susceptible to induced noise.

'Current loops' are traditionally considered a good means of transferring signals up to a few hundred metres. In a current loop, the transducer only allows current to flow in proportion to the input to the transducer. Typically the current will vary from 4 to 20 milliamperes (an industry standard). Currents outside this range are used to indicate fault conditions. However, the performance of a current loop relies on there being no leakage whatsoever in the circuit. On one recent occasion it was found that the angle of a massive bridge abutment apparently changed each time it rained. A short circuit between the sensor circuitry and its case provided a leakage path to earth when the masonry was wet.

Sensors which can be driven using a precision high frequency AC voltage have an intrinsic merit when used with signal conditioning which incorporates 'synchronous detection circuitry'. The signal conditioning at the receiving device can distinguish that part of the signal which is synchronised with the drive and reject anything else as noise.

Frequency modulation offers a means by which signals can be transferred through cable over considerable distances without loss of accuracy, and generally with a very low susceptibility to interference.

5.2.4 Interconnection

Generally, the sensor is mounted at the position where the measurement is to be made, except when one of the remotely mounted non-contacting measurement technologies is employed. The reading will usually be made at some remote and more convenient location. The two points will require interconnection.

In the past, interconnection has been achieved mostly using a signal cable which carries the electrical analogue of the parameter to the point measurement. In recent years other transmission techniques and media have become more common. These include analogue systems incorporating (typically radio) telemetry links, and digital systems using wire, fibre optic and telemetry links.

In the majority of analogue signal networks each analogue signal is passed through an associated cable to the monitoring location. Cables may be bundled together for part of their length and, when appropriate, multicore cables will be used to carry the signal from several sensors or transducers to the measurement point. Occasionally signals from several sensors will be combined or multiplexed and sent together down a single signal cable to the measurement point where the combined reading is measured or the signal is de-multiplexed. Analogue signal networks commonly take the form of a hub with spokes. When signals are combined or multiplexed the network is more akin to a trunk with roots and branches, with one root for each branch.

These styles of network can involve considerable lengths of cable if the sensors are widely distributed or the hub cannot be located centrally. The cost of installing cables grows rapidly as the scope of the system is increased and soon becomes a significant part of the overall cost of the system (in some instances more than half the cost). Multiplexing and demultiplexing is also expensive and will only be effective in a limited range of circumstances.

Cabling costs and inconvenience can be avoided by using short range radio links if the sensors, plus any associated signal conditioning and the telemetry link, can be powered locally (from an existing power supply or batteries). However, the radio links are themselves expensive and they also may reduce the overall accuracy of the system.

An alternative to the 'hub and spoke' style of network is provided by more recently developed digital systems. These use a shared data network to provide a means by which digitally encoded data is passed from the sensor to the data-acquisition unit. The data-acquisition unit will typically consist of a computer equipped with software which will enable readings to be obtained from individual sensors or groups of sensors (if required at very high speed) by requesting and receiving digitally encoded data.

Digital systems of this type offer the possibility of greatly reducing the number, extent and therefore cost of the interconnections. The resulting network can take the form of a 'daisy-chain' with one cable or link connecting each sensor or group of sensors in turn or it can have a trunk, branch, and twig structure. The links will be routed to visit all the sensors or each group of sensors using the shortest or most convenient route. Where 'hard' links cannot be employed, digital telemetry links can be incorporated and ultimately cableless systems can be obtained if power can be sourced or provided locally.

The penalty incurred by such systems is one of increased sophistication and escalating cost. Each sensor has to be provided with the means to respond to a request for data with an accurate report of the current data value. This is achieved using microprocessors and analogue-to-digital convertors. The microprocessor is programmed to monitor the communicating link for requests for data issued by the controlling computer. When a request is received, either an analogue-to-digital conversion is carried out and the outcome is reported, or the result of a previous conversion is reported. More sophisticated equipment continuously monitors the sensor output and filters, converts, and records the data (or reports new or historic data) as required. The falling real cost of such equipment means that the threshold at which this technique is appropriate is reducing with time. Ultimately it can be expected that analogue data will rarely be transmitted over any appreciable distance.

There are also composite systems which reduce the amount of cabling involved in purely analogue systems by using a series of outstations which are all connected to each other by a digital data transmission network and which each have a number of sensors connected to them using more conventional analogue signal transmission techniques. Transducers and sensors are grouped together on the basis of physical proximity. Each one is treated as one 'spoke' on a local 'hub'. The 'hubs' are then connected via a digital link to the computer which controls the system. Systems of this type have a great advantage in that they are inherently modular and can be extended in scope by the addition of further 'hubs'. The hardware cost is therefore proportional to the scope and size of the system.

A system which is specifically intended for monitoring in a construction industry context, the distributed data-acquisition system (or DDS), is outlined here. This system consists of modules which each provide four inputs. Any number of modules, known as distributed data-acquisition modules (or DDMs), can be linked together using the appropriate transmission medium. The whole system is automatically controlled by a specially adapted computer which can be located at any point in the DDM network or at the end of a modem or telemetry link. Although this represents a fairly sophisticated system, it is extremely cost-effective when used in appropriate circumstances. Smaller systems which can run autonomously use a specially adapted version of an inexpensive hand-held computer. Larger systems benefit from the use of a suitable personal computer. The interconnection scheme reduces installation costs dramatically, with the result that most of the cost of such systems

Figure 5.2 A distributed data-acquisition system can be expanded indefinitely by daisy-chaining data-acquisition modules.

can be attributed to the purchase of system performance rather than system installation. In contrast, some 'hub and spoke' style systems require a very high proportion of the installed cost to be spent on interconnection, with commensurately little spent on the more critical area of hardware and software investment.

Figure 5.3 The controlling computer can be closely coupled to the data-acquisition module or remote from it. Digital communications are unaffected by the distance involved.

The data transmission technique and the transmission medium will affect the style of monitoring that can be carried out. If a true analogue of the input to the sensor or transducer is taken to the monitoring position, then it can be recorded using analogue devices and its analogue nature preserved. If a digital transmission technique is used then the data transferred will always consist of discrete packages of information. A high enough data rate in a digital system will allow a good (or even excellent) approximation of an analogue output to be obtained (subject of course to careful consideration of the potential for 'aliasing').

Conventional copper conductor signal cables constitute the predominant means for transferring both analogue and digital information. Such cables are available in a wide range of styles to suit different environments: at one extreme are light PVC sheathed cables of small overall diameter and at the other, heavy wire layer armoured cables which are suitable for use in extremely hostile environments.

Fibre optic links provide an alternative for digital signal transmission combining the benefits of complete electrical isolation between different parts of a system with very high data transmission capacities and small physical size. Electrical isolation can be an extremely important consideration in system design. It can be crucial to the safety of the operator: in one example a welder struck an arc on a bracket supporting a cabinet containing some monitoring equipment. Adequate isolation protected the remainder of the system and the operator though unfortunately not the contents of the cabinet. Isolation can also be very important where earth potentials vary around a structure as this may interfere with the functioning and performance of the monitoring system.

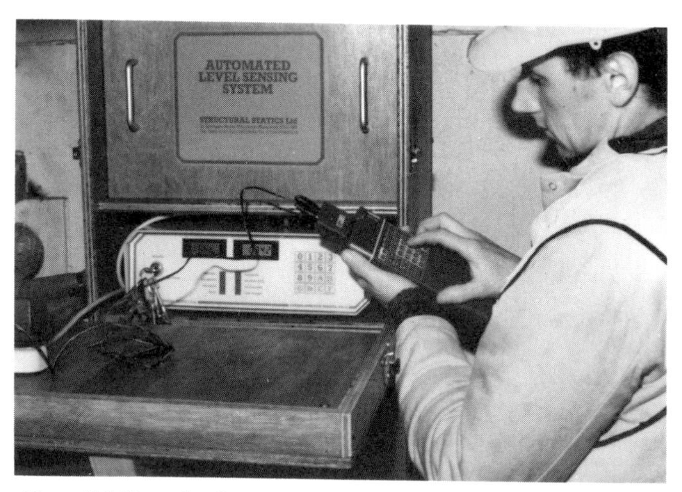

Figure 5.4 Down-loading data from an automated level sensing system.

Telemetry links can transfer both analogue and digital data over considerable distances. If there is to be no physical connection between the different parts of a system then each part must be locally powered. Low energy systems can be powered from batteries, and these can be replenished by solar cell arrays.

The use of modems and conventional telecommunications provides the possibility of systems being controlled from far afield. For instance the distributed data-acquisition system referred to earlier can be left to work autonomously and the data generated can be instantly down-loaded via telephone. Equally the DDS can be operated through the telephone system thus obviating the requirement for a computer on site. Satellite and 'meteor burst' communications systems can be used to establish links between any two locations on the earth.

5.2.5 Data-acquisition equipment

At the monitoring location the required data must be captured, recorded, and displayed. The appropriate options available will depend on the manner of the medium for transmission. Analogue signals can either be handled as analogues or converted to digital. Digital data will generally be handled in digital form until it is finally displayed in a quasi-analogue graphical format for ease of assimilation.

The simplest form of analogue display is the moving needle of a dial gauge with a graduated scale. In most instances this is all that is required. Access to the data is instantaneous and trends can be observed with ease. However, such recording necessarily involves the operator unless the display incorporates some form of secondary signal output to a recording device.

Groundwater levels are monitored closely in the City of London on account of the implications of groundwater changes for the many historic buildings which are founded at relatively high level. In many instances the readings are regularly read from dials and recorded manually. In one instance a sudden change of water level was initially attributed to equipment failure, but further investigation revealed that the equipment was working properly and in fact a nearby water main had failed. The system is simple, cheap and effective, although it does depend on fairly lengthy manual record keeping and data transfer methods. It is now possible to install single channel monitoring systems to carry out this type of task relatively cheaply. These systems are capable of gathering several years' worth of data which can swiftly be downloaded to a portable computer and rapidly presented in graphical and tabular formats.

Microprocessor-based gauges are available which provide both an analogue and numeric indication of the value of an input scaled to the appropriate units. Some offer a graphical output which enables the trend to be immediately visible. Many provide both high and low alarm setting capabilities, opening and closing relay contacts or a 'logic' output to another device. Some are also capable of reporting a regularly updated result to an attached computer, printer or other suitable recording device.

Liquid crystal displays and cathode ray tubes (CRTs) offer huge variety in the way analogues of the parameter of interest can be displayed. These range upwards from relatively simple simulations and variations on the theme of the dial gauge. Oscilloscopes offer a means by which (part of) the history of the output from a transducer can be presented graphically and new information can be displayed in real-time. Oscilloscopes (digital) with recording facilities can be used to replay the recorded information, at variable speeds.

Computer- and microprocessor-based systems can offer sophisticated mimic displays. When the information is available in digital form it even becomes relatively simple to produce graphical representations of the building in question which mimic the movements observed, normally in an appropriately exaggerated fashion.

Recording analogue signals from transducers and sensors is straightforward. The two most common means are the pen-plotter/chart recorder and magnetic tape. The main drawback is that the amount of the medium used is directly proportional to the period of time over which monitoring is carried out coupled with the speed at which the recording medium is moved. This can result in very large volumes of tape or paper being consumed and in associated data handling problems, particularly when the incidence of an event of interest is infrequent and of short duration. Equipment is available which can be programmed to isolate and record these events: a threshold is set which defines when recording is to start. Some types of equipment cease recording after a certain time has elapsed following the signal recrossing the preset threshold. Others provide the facility to set another threshold to define when recording should stop. Better equipment buffers or stores a certain amount of information so that the recording device can be given time to start working. The most sophisticated equipment will buffer a fraction of a second, several minutes or even hours of information so that the period immediately preceding the triggering of recording can be recorded as well. This style of recording is particularly valuable for isolating nuisance phenomena such as intermittently occurring vibration or excessive noise in a building.

Pen-based chart recorders suffer inherent limitations regarding frequency response. The mechanism controlling the pen position on the paper may be incapable of moving the pen far enough and fast enough to record the input accurately. High frequency signals will therefore be attenuated and important information lost. This can be a very important consideration in the choice of a suitable recording equipment for faster rate phenomena. Recorders which use heat sensitive paper as the recording medium can offer better frequency response characteristics, while recorders using optical and electrostatic recording techniques can offer considerable accuracy at frequencies higher than those generally applicable to building structures.

Hybrid chart recorders accept analogue inputs which they digitise. The digital information is then presented on paper numerically and/or graphically. Most of these devices employ dot-matrix printing technology to transfer the information to the recording paper and most are capable of accepting multiple inputs. Some offer as few as four input channels and others accommodate 32 or more inputs. The greater the number of inputs, the slower is the maximum recording speed. The more crowded the paper becomes, given the limited width of the paper, the more difficult it is to interpret the data. In practice the relatively slow process of transferring the data to the paper limits the frequency response of these devices and renders them suitable only for monitoring fairly steady inputs.

The hybrid chart recorder is an intermediate between the all analogue and the predominantly digital system. Digital systems and equipment fall into two broad and reasonably distinct groups: dedicated data-acquisition equipment and systems plus equipment which are in essence peripheral devices for general purpose computers. The latter includes the vast proliferation of analogue interfaces for all types of computers ranging from portables to mainframes.

Most dedicated data-acquisition equipment has been designed and built to serve the interest of a particular market. Many are targeted at the industrial process control market. Scientific research and development have particular needs which are met by equipment tailored to suit the laboratory environment. Meteorological and agricultural instrumentation requirements have also led to the creation of a wide range of data-logging systems and equipment capable of autonomous operation in hostile environments: these are often battery rather than mains powered, a feature which can be essential in some circumstances. Unfortunately environmental data-logging is rarely carried out with any great precision and much of this equipment is unsuitable for structural monitoring.

Stand alone data-loggers fall into two categories depending on whether they are intended for attended or autonomous operation. For autonomous operation they are generally 'blank faced' and do not provide any particularly sophisticated means for displaying acquired data. The data they record will be transferred either to a computer or dumped to a printer, a dedicated data transfer device, or a removable recording medium. Data acquisition equipment which is intended for hands-on use will offer a much more sophisticated means of presenting the data to the user, both as it is acquired and from record. Typically the data will be presented to the user on a screen and there will also be facilities for presenting data using a printer or plotter. Depending very much on the market for which the device in question is intended, it will operate with a modest range of sensors and transducers without the need for any additional signal conditioning.

Temperature and temperature control are a major preoccupation in industrial process control, and there is a formidable range of devices available which can be coupled directly to all forms of standard temperature sensors. These sensors include a wide range of resistance temperature detectors (RTDs), thermocouples, and more recently semiconductor thermistor temperature sensors. RTDs offer reasonable accuracy over wide temperature ranges, thermocouples offer very low thermal mass and consequently very fast responses to temperature change, and thermistors offer good, reliable performance coupled with ease of use—provided the equipment can cope with a logarithmic polynomial transfer function. Some of this data-logging equipment also offers the ability to accept other (high level) voltage inputs and to scale the reported output to the units required by the operator. This affords the opportunity for their use in structure monitoring with appropriate combinations of sensor/transducer and signal conditioning. A subset of this group of devices is able to be coupled to quarter, half and full Wheatstone bridge circuits. This feature enormously increases the range of parameters which can be monitored without the need for additional signal conditioning, as a great variety of parameters can be measured using sensors which are based on, or behave like quarter, half or full bridge circuits.

From the perspective of structure monitoring, the capacity to accept high level inputs (e.g. volts, as opposed to millivolts), and to scale the input to provide output in the desired engineering units, is crucial. The ability to use rather more sophisticated 'transfer functions' can be very valuable.

Most stand-alone logging devices offer eight or more input channels. A few offer only one to four channels but some offer (normally by the addition of further hardware) hundreds of inputs. It should be appreciated that the additional channels are normally achieved by multiplexing the inputs. The data-acquisition equipment is connected to each channel or bank of channels in turn. The switching process will take anything between fractions of a millisecond and seconds according to the sophistication and accuracy of the device. This can dictate the choice of equipment where faster rate phenomena are to be monitored.

The upper end of the range of dedicated data-acquisition equipment and systems is represented by computers specifically adapted to the task. Subject only to their fulfilling accuracy and resolution requirements, their potential usefulness is limited only by the scope of the software which controls their data-acquisition, handling, recording, display and presentation functions. Two styles exist which reflect the markets at which they are targeted. The first assumes that all measurement points are within a reasonably short distance of one central measurement location. They require a star or hub and spoke style of interconnection. The second assumes a wider physical distribution of measurement points and is typically based on daisy-chain interconnections between remote hubs with spokes. The latter is primarily intended for supervisory, control and data-acquisition (SCADA) applications in industry involving both monitoring and control.

Virtually any computer can be converted into a data-acquisition tool by the addition of hardware which will convert varying signals into digital values coupled with software which drives the hardware and provides the necessary data capture, display and recording functions.

There is now an immense range of hardware and associated software available which can transform modest computers into powerful data-acquisition tools. Some of this equipment consists of stand-alone hardware with which the host computer interacts via a standard communications port. Other equipment interfaces directly to the computer's internal bus. In the case of IBM PCs and compatibles, this is frequently achieved by slotting a card into an expansion slot. Internally mounted hardware of this sort can range in capacity from one to sixteen channels. Higher channel counts usually require external hardware for cable terminations. Externally located equipment in some cases is capable of virtually unlimited expansion by duplication or replication. The range which is available continues to grow at a great rate, it is therefore not possible to define or characterise all possible options. The following observations are necessarily generalisations:

- A computer's speed of operation is extremely fast. Accurate measurements are not easily made at great speed without commensurate sophistication and expense. Many manufacturers of data-acquisition hardware for computers have tended to opt for speed rather than accuracy. However, in many instances structure monitoring tasks require high accuracy.
- A high proportion of this type of monitoring equipment is only really intended for relatively short-term monitoring programmes usually lasting only a few minutes.
- Attaining high resolution is expensive. Many devices offer only eight-bit resolution: the maximum permissible range of the input is divided into 256 steps. Taking account of such factors as the necessary excess of the range of the sensor over the actual input range and mismatches, this is frequently inadequate. Twelve-bit resolution or better is usually required for structure monitoring tasks.
- Many data-acquisition add-ons for computers have been designed only for use in the laboratory or workbench environment. They assume that all the associated equipment will be at or near the same voltage/potential. They do not provide adequate electrical isolation between the computer and the sensors for safe use in the typical structure monitoring environment. While the risk to the user may be small, the risk to the computer is likely to be unacceptable. Monitoring systems frequently extend to parts of buildings in which there are three-phase supplies or to parts which are on different phases of a three-phase main supply.

5.2.6 Display

Computer-based data-acquisition systems and most dedicated data-acquisition equipment which is intended for 'hands-on' use are capable of dis-

Figure 5.5 An industrial PC can gather data 24 hours a day, 365 days a year. It also provides the means to process and review the data obtained.

playing the data as, or soon after, the data is obtained. The available options for data display will depend on the sophistication of the software which is provided. Inevitably these options will reflect the perceived requirements of the target market for the device or system in question. Monitoring building structures more often than not imposes requirements which are specific to a particular task. The options offered by standard software frequently satisfy only the most straightforward requirements. Text based displays of data do little to reveal trends and even tend to conceal them. Graphical data presentation may not offer adequate resolution unless the output can be scaled and offset appropriately because of the limited screen resolution. As a result it is frequently more effective to think in terms of separating the data-acquisition and presentation tasks to overcome such limitations. One of the most effective ways of doing this is to make use of standard database, spreadsheet and graphical presentation software to process data which has previously been recorded on magnetic media. This separation is readily achieved when the monitoring task is carried out automatically and autonomously, and the data obtained can be taken away for appraisal and analysis.

Economic constraints frequently preclude putting anything other than a simple, if robust, matrix-printer on site. These are primarily intended for text printing tasks and are painfully slow at producing low quality graphical output. Laser printers driven by suitable software produce higher quality output at a more acceptable speed. Plotters driven by good software can achieve similar quality of output at comparable or even greater speeds.

5.2.7 Recording

In the early days of automatic and autonomous structure monitoring, before the advent of relatively cheap mass storage devices, it was quite usual to record data on paper as it was obtained. This was cumbersome and inefficient. Such systems would produce vast quantities of data and paper, and the data would need to be re-keyed if any form of computer based analysis was to be carried out. Nowadays paper should be used only for data presentation or archive purposes if absolutely vital. With the exception of the chart recorder based monitoring system, paper should be avoided as a recording medium. Failure to do so nullifies one of the great advantages of an automated monitoring system; its capacity to provide data in a readily handled and manipulated format.

Hard and floppy disc systems are presently the best available storage media. The former offer enormous capacity, the latter ready portability, and both systems offer reliability and integrity. All the usual issues pertaining to the use of magnetic storage media are relevant. All magnetic media based records have to be handled in a disciplined fashion if data are not to be lost, overwritten, wilfully or unwittingly erased. A further word of caution: one system built used an industrial PC as the host device and as the operating temperature range was adequate for the conditions in which it was going to be used, no problems were anticipated. However, the system proved to be unreliable with the advent of summer. The temperature was well within the operating temperature range but it transpired that thermal expansion and contraction of the floppy disc was at fault. Until the computer was moved to a thermally controlled environment, summer and winter discs had to be employed.

Magnetic tape cartridges provide an alternative cheap, large volume digital data storage medium. They suffer the disadvantage of being linear media, which makes data-access and recovery a more time-consuming process.

Solid-state recording of data offers many attractions. Physical robustness and speed of access are two in particular. However, unless the memory used is either protected by an uninterruptible power supply of sufficient capacity or it is otherwise secure against power supply failures (even with battery powered equipment), there will remain a risk of losing data. 'Write once, read many' forms of solid-state memory offer very high levels of data security. These include E and EE PROM memory and optical systems. The former can be quite cheap but the latter is presently expensive.

5.2.8 Recording strategies

The unassailable advantage of properly designed automatic and autonomous monitoring systems is the abundance and assured quality of the data. Even

Figure 5.6 Reviewing and processing data on a PC in the comfort of the office.

using quite powerful desk-top computers and sophisticated software, manipulating the abundant data generated can quickly become time-consuming and laborious. A strategy is therefore required which reduces the data retained to an acceptable minimum but which nevertheless retains all data that are pertinent.

In an extended programme of measurement it will be appropriate in the very early stages to gather much more data than will be necessary as the programme advances. It is only by doing this that it will be possible to accurately gauge the context in which each measurement is made. Automatic monitoring systems are generally capable of obtaining virtually instantaneous readings. The response to change of some sensing technologies is so fast that instantaneous values for quite high speed cyclical phenomenon can be measured when what is actually sought is a mean value.

A system which was designed to be capable of furnishing a snapshot of the inclinations of the towers and the changes in level of the deck of a long span bridge proved, when first commissioned, to be too fast. The contractor required accurate information about the average position of the towers and deck. Adjusting the way the sensor outputs were filtered produced the desired effect.

All structures are subject to diurnal heating and cooling. The scale of the resulting movements may not be significant in a relatively massive masonry or concrete structure, though easily capable of damaging brittle finishes if provision for expansion has not been made. In steel and steel composite structures, temperature changes can cause movements capable of completely masking the underlying trend which is the actual cause for concern.

Knowledge of diurnal and seasonal (and occasionally lunar) cycles may be valuable in the context of automated monitoring. These cycles may be significant where the extent of the movement has to be accommodated in a proposed modification of the structure or where these movements are themselves the cause of the problem. However, they may simply be the background to the real concern. In many cases, once quantified they cease to be of concern and provided that the phenomenon being monitored is occurring significantly more slowly than the cycle rate, the rate at which data are obtained can be reduced accordingly.

Transient phenomena pose rather different problems. Frequently it will be of particular interest to have as much data as possible in the period immediately preceding and following the occurrence of the event. Means are required for maintaining a 'rolling history' which is only recorded when a trigger is activated. There are available (pseudo-)analogue and (wholly) digital data-acquisition systems which may be set to function in this manner. Both types of system function in the same manner. Data are 'captured' digitally at a very high rate, and recorded in memory. When the memory is full the oldest data are then over-written. When the trigger is activated the contents of the memory are either printed, displayed or recorded.

5.3 Sensors and transducers for a variety of applications

5.3.1 Linear displacement

One of the most common applications for automated structure monitoring equipment is in monitoring the most obvious evidence of distress such as fractured concrete, stone or brickwork. There are several reasons why monitoring may be required. The break may be attributable to one or a combination of possible processes. Monitoring may be called for to confirm an obvious mechanism, to reveal a less obvious one, or to show that the one originally responsible for the damage is no longer of concern.

Close monitoring of the structure may allow more appropriate remedial works to be designed, and very expensive remedial works to be deferred: it may even prove that remedial works are unnecessary. In some instances close automated monitoring has shown that the specified remedy would have been wholly inappropriate. Where a structure has broken to provide necessary freedom of movement it may not be appropriate to attempt to constrain that movement. Equally, making cracks good by filling them with strong materials can cause considerable damage when the crack tries to move in response to weather or temperature changes. In extreme cases, immediate knowledge of further movement may be sought so that the use of a building may be continued while repairs are carried out.

In the majority of situations movements at a crack, gap or joint can be monitored using devices which bridge the opening. The device used must provide freedom of movement over the whole range of movement and make

Figure 5.7 A displacement transducer installed over a construction joint. A length of foam tube provides some mechanical and weather protection.

due allowance for the unexpected. It must offer precision appropriate to the smallest movement which is of interest. These can be conflicting aims as a range of movement which is adequate to cover the most extreme movement which can be anticipated may not offer sufficient accuracy or resolution to measure precisely enough the smallest increments of movement which are of interest. However, it is often possible to overcome these conflicts.

Displacement sensing is fortunately such a common requirement in industry and elsewhere that the range of devices available is immense, and even quite exceptional requirements can be met with standard devices. There are a number of quite different sensing technologies available each of which have relative merits and demerits. The well-developed market for displacement transducers also means that price, features and performance are interrelated in a reasonably intelligible fashion.

The simplest displacement sensing technology is the linear potentiometer. The principle employed is the potentiometric divider: a contact is caused to move along a conductive track by the movement at the crack or joint. Applying a precise voltage across the length of the track and equally precise measurement of the voltage at the contact allows the position or change of position to be ascertained, as the electrical potential depends on the contact's position in relation to the ends of the track. The accuracy of the measurement is obviously dependent on the quality of construction of the sensor used: the voltage must vary in a uniform (or predictable) fashion along the length of the track and the mechanism must not distort in use. Also, only very little current can flow through the contact to the voltage measuring device if the accuracy of the system is to be maintained.

Wear and tear can affect the performance of potentiometric devices when they are used for extended periods. If vibration or cyclical movement is present this will ultimately cause wear on the track and the contact which will affect the electrical characteristics of the sensor (and therefore its accuracy). In some instances the track material will age or degrade, which can affect the output and accuracy over longer periods.

Potentiometric displacement transducers range from relatively cheap and low precision plastic bodied devices, to high quality, high precision units incorporating such features as oil bath lubrication, dust sealing and even submersibility.

A wide variety of different mounting arrangements are also available. For a simple crack width measurement a mounting arrangement which uses spherical universal joints at both ends of the transducer provides positive control of the position of the contact and yet will accommodate off-axis movements: movement at cracks and joints is seldom conveniently limited to the plane parallel to the surface.

Precisely-ground mounting pins can be made to provide a very close tolerance fit in locations in the spherical joints. These mounting pins can be attached to the structure by embedding them in a suitable epoxy material or using intermediate mounting bracketry. The more direct the mounting arrangement, the less scope there is for introducing inaccuracy. Differential heating of even a small bracket (for example by sunlight) can cause unacceptable deformations when higher precision is sought. An enclosure can circumvent this problem and also can provide a more aesthetically acceptable appearance to the installation.

Some transducers offer spring-loaded gauging heads which can be caused to bear onto a small (flat) target plate. This can be very useful when measuring movement at a crack or joint in the corner of a re-entrant angle. However, spring-loading does not provide positive control of the contact position and care must be taken to prevent anything finding its way between the target surface and the gauging head.

Electrical simplicity is the great merit of the potentiometric transducer, although it does have some detractions. The excitation voltage must be applied to the track and not to the wires attached to the end of the track and the voltage must be measured at the contact if accuracy is to be maintained. Voltage drops along the length of cables can appreciably diminish the accuracy. In a dirty or noisy electrical environment induced voltage can wreak havoc. The solution is the use of signal conditioning in very close proximity to the transducer. Modules are available which will convert potentiometric transducers into current loop devices. These apply a precise voltage to the track, monitor the returned signal and allow a current in the range of 4 to 20 milliamperes to flow around the current loop. A precision resistor in the loop can then be used to drop a voltage to suit the input of the monitor device. Some sensors which use the potentiometric principle have signal conditioning of this type built in.

Another family of devices employs strain gauges to obtain or provide an analogue of displacement. The movement of interest is made to bend a sensing element. This gives rise to strain on the surface of the sensing element which is monitored using strain gauges. The strain gauges are arranged in a bridge so that a voltage proportional to the applied voltage and the displacement can be measured. This is the principle employed in extensometers commonly used in testing laboratories. There are also transducers which employ a variation of this technique. In these the movement of a conical plunger causes a number of fingers to bend and the resulting strain in the fingers provides the analogue of displacement. These devices are more robust than conventional extensometers and can provide far greater ranges of movement.

The use of Wheatstone bridge circuits is commonplace in all forms of sensing systems. It is frequently used when temperature and pressure measurements are made because much of the equipment available is designed to make measurements using bridge circuits. Displacement sensors which use the same principle, therefore, have the merit of compatibility with very many other sensors, sensing and data-acquisition systems.

The output from a bridge circuit is low-level, typically expressed in terms of millivolts/volt applied to the bridge. Like potential dividers, bridge circuits are also relatively high impedance signal sources: significant currents cannot flow without degrading the accuracy of the signal. These factors require careful consideration or the provision of additional signal conditioning if accuracy is to be ensured. Signal conditioning modules are relatively easily obtained but they usually require a detailed understanding of their use as they will normally need adjustment to obtain optimum performance.

Linear variable differential transformers or LVDTs are devices which are particularly well suited to displacement monitoring in building structures. The principle they employ is most readily described in terms of two transformers set up to work against each other. The windings of the transformers are enclosed within a housing. A metal slug or armature is displaced by the movement of the structure, enhancing the inductive coupling of one transformer and degrading the inductive coupling of the other. This results in a signal which varies with the displacement either side of a null point at which (nominally) no signal is returned.

These devices offer a number of benefits which are relevant to the applications considered here: they can offer very high precision when it is required; they are mechanically and electrically robust; and they can readily be made waterproof and (when required) submersible. In addition, because the coupling is inductive there is no mechanical wear and tear on the measurement system in use, and they therefore offer the advantages of long life and enduring precision.

Their disadvantage is a requirement for a particular type of excitation and signal conditioning. Many devices which use the LVDT principle have this built in. These DC LVDTs (as they are known) require a regulated DC voltage supply and typically return a DC voltage (varying with displacement). However, the electronics of many DC devices will not accommodate the ambient temperature range to which they can be exposed in structure monitoring or do so only with much degraded accuracy. On the other hand AC devices can accommodate a much wider temperature range without significant loss of accuracy. Their use with appropriate styles of signal conditioning also means that long cable runs can be employed without loss of performance (but normally subject to *in-situ* calibration). They are also not prone to electrical interference.

There is another family of displacement sensors which use changes in capacitance caused by displacement to obtain an analogue of that displacement. These translate displacement of a gauging head or plunger into small changes in the gap between the electrodes of a capacitor. The changes in the capacitance (of the capacitor) which result are then converted to a high level output signal. These devices can offer extremely high precision. They have primarily been developed for monitoring and controlling machine tools. Accuracies can be achieved which are expressed in terms of parts of a thousandth of a millimetre. Unfortunately they are extremely expensive.

5.3.2 Displacement about several axes

There are many instances where the movements of interest must be resolved on two or more axes. There have also been occasions when transducers have all but failed to detect appreciable movements because of incorrect alignment on installation.

Resolving movement on two axes in a plane parallel to a surface can be achieved by mounting two transducers so that they are on orthogonal axes. At a straight crack or a construction joint this will involve additional transducer mounting hardware so that one of the sensors can be mounted parallel to the crack or joint and one across it (or one vertical and one horizontal). Provided that the movements are small in relation to the length of the transducer the output in each will represent the movement on each of the two respective axes.

An alternative is to mount the transducers in a V-shaped array. Two mounting pins are mounted on one side of the crack and a single target pin is mounted on the other. Two transducers each sense the movement of the common pin with respect to the base of the triangle. Simple trigonometry is then used to resolve the magnitude and direction of a relative movement. This arrangement is neat, simple to install and has been used very successfully in a wide variety of situations.

A recent exercise concerned a situation where rotation as well as translation was of interest. A displacement and rotation monitoring unit (DRMU) was developed which uses three displacement transducers to monitor movement across the width of a construction joint or crack and along its length, and

Figure 5.8 Displacement transducers installed in a V-shaped array so that the extent and direction of relative movement can be resolved.

also detects rotation of the fabric on one side of the crack/joint with respect to the other side. The DRMU is mounted on four mounting pins which are set in pairs into the fabric on either side of the crack using a precision installation jig. The transducers are fixed to profiled plates which fit onto each pair of pins and gauge off the complementary plate.

In addition, there are situations where monitoring must also be carried out on a third axis which is at right angles to the surface. In a re-entrant angle this presents little difficulty. At a joint/crack in a planar surface some form of stand-off bracket will be required. This must of course be adequately strong

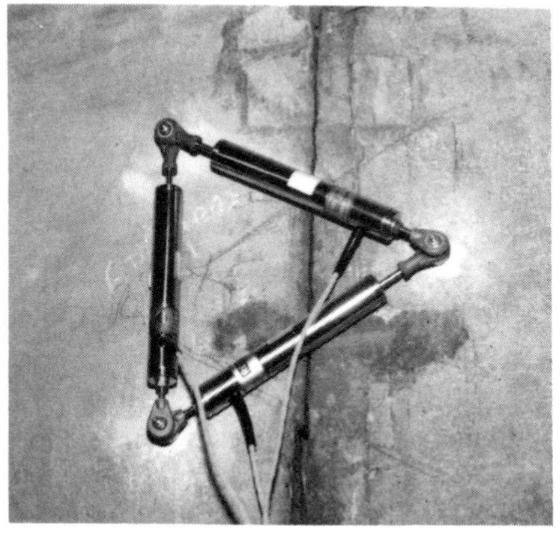

Figure 5.9 A third displacement transducer was installed at this location to allow the stability and temperature sensitivity of the measurement system to be verified.

Figure 5.10 Three displacement transducers installed in a re-entrant angle to permit measurements to be made on each of three orthogonal axes.

and stiff. Depending on the size of the bracket and the scale of movement which is of interest, there is some risk that small movements may be hidden by thermal movement in the bracketry. A temperature sensor mounted in the vicinity can be used to make corrections. Enclosing the sensor and the mounting hardware so that differences in temperature are reduced to a minimum will also help. Strong sunlight on a steel bracket, or part of it, can cause a great deal of movement.

Figure 5.11 An array of three transducers. Measurement of movement on the third axis required a suitably stiff stand-off bracket.

5.3.3 *Contactless displacement*

There are instances where it is not appropriate to bridge the gap across a crack or a joint. This may be because there is known to be movement along the line of the crack or gap as well as at right angles to it. It may be undesirable to introduce, or risk introducing, restraint. Monitoring the movement caused by wind loading of a curtain wall with respect to the enclosed structure recently imposed this restriction. Failure of the seals was attributed to excessive movement and the designers were anxious to establish what character of movement was normal, as well as what happened in exceptional circumstances.

Varieties of LVDT are available which can measure across a gap of a few to a few tens of millimetres. In common with the more orthodox forms of LVDT they offer high precision (repeatability) and are robust. The output is normally highly non-linear and has to be corrected by reference to a calibration table or by approximation over small parts of the range.

Alternative devices use an eddy current effect to sense the position of an adjacent conductive target surface over ranges of a few millimetres up to 250 millimetres. A high frequency signal in a coil is damped or attenuated by the eddy current it causes in an adjacent conductive surface. The attenuation obeys an inverse square law. Signal conditioning measures this effect and provides an analogue output which is proportional to the gap. In the example of the curtain wall a device of this type was used to monitor changes in the position of a thin foil which was pasted to the inside of a glass panel.

Measurements can also be made across a gap using a variety of optical devices. On the whole these consist of self-contained units or subsystems which require some form of power supply and return a voltage which is an analogue of the gap. The principles used vary from device to device. The intensity of the reflected light and the size and position of the reflected image are each used to determine the position of the target.

Transducers which employ the principles of the marine echo-sounder are also available for monitoring across larger gaps. Resolution and accuracy frequently present difficulties. These devices rely on a crisp echo. Their performance can be degraded where there is noise in the background, particularly when this is caused by spurious reflections of the signal transmitted by the transducer.

5.3.4 *Long baselines*

There are instances where the objects whose relative movements are of interest or concern are not conveniently located immediately adjacent to each other. Quite frequently the structural element which is thought or known to be moving is some distance from a component that can reliably be assumed to be stationary. In these cases, when continuous monitoring is called for, some means is required to extend the baseline of the measurement. However, extending the baseline will invariably increase the temperature sensitivity of the measurement necessitating additional compensation and/or correction.

The transducers which can be deployed to measure across gaps and cracks can frequently be modified by the addition of extension pieces so that they can measure small changes on a fairly long baseline without loss of accuracy provided that appropriate corrections are applied if compensation is not built in.

When spherical bearings are attached to the transducer using the male or female threads provided by the manufacturer, it is relatively easy to extend the body of the transducer using an extension piece which provides a male thread at one end and a female thread at the other. Using Invar for the extension piece will reduce the temperature sensitivity of the assembly. In practice, temperature correction is readily achieved if a temperature sensor is sited nearby or, better still, built into the extension piece. Extension pieces and rods can be used for baselines up to several metres in length. A length of scaffold pole can be most effective in appropriate circumstances.

At greater distances, and particularly when the measurement is an horizontal one, tensioned wires provide a rather more effective means for achieving the measurement. The prerequisite for the effective use of wires is a means to apply a constant (or precisely predictable) tension to the wire at all times regardless of relative movement of the end points. The wire itself is in effect a long spring, and a change in tension will produce a proportional change in length.

A constant tension is most readily applied by a dead weight. Springs invariably introduce a small measure of uncertainty. Most other means of applying tension involve too much friction or stiction.

A dead weight applied by a lever with a knife-edge bearing system can be used very effectively if the components can be sized to avoid significant geometry changes over the combined range of actual and thermal movement. In a controlled thermal environment, particularly below ground, this can be readily achieved even when the baseline is tens of metres in length. In normal outdoor ambient temperatures this is less practicable.

Pulley-based systems offer much greater ranges of movement within a smaller envelope although these have inherent difficulties associated with the pulley bearings. A recently developed tensioned wire reference unit uses a sophisticated frictionless bearing system to support a pulley wheel to achieve an effective compromise. The frictionless bearing is actually a low spring rate flexure which allows the pulley to rotate freely. When relative movement takes place the flexure is wound or unwound depending on the direction of movement. The spring rate is so low that the increase or decrease in wire tension is insignificant.

Over longer baselines it is frequently not possible to consider that the whole of the measurement system is within an isothermal environment, even when

Figure 5.12 A system using tensioned wires allowed the consequences of releasing and removing temporary tie bars to be monitored closely.

Figure 5.13 The tensioning unit applies a uniform load to the Invar wire. The relative movement of the tensioner and the wire end is monitored.

Figure 5.14 This tensioned wire system was installed at St Paul's Cathedral. Using parallel wires made from different materials allows temperature effects to be accurately corrected.

enclosed in a duct or casing. Temperature compensation then becomes more difficult. This was overcome at St Paul's Cathedral where parts of a long baseline tensioned wire system passed through open air and the remainder was within the roof space. Two parallel systems were installed. Each used different wires with known relative thermal coefficients. The wires share a duct and are therefore exposed to a common thermal environment along their length. Combining the output from the sensors permits full temperature correction. The average temperature of the wire is also established as a by-product of the calculation.

When the baseline for the measurement extended beyond a few tens of metres and particularly out to open air, recent adaptations of traditional surveying techniques (similar to those in section 2.4) proved successful, particularly those which offered interfaces to computer-based data-logging systems. This is a field which is presently being led by the requirements of major civil engineering projects though the techniques and technology will eventually be applied to building structures. However, costs are at present very high when equipment is installed for continuous use. For the time being this may be considered a useful adjunct to automated monitoring.

5.3.5 Inclinometers

The concern so far has been with means of measuring the relative movement of one object or point with respect to another along an axis which is the straight line between the two measurement points. Automated structure monitoring frequently includes requirements for measurements of changes of inclination. This change in inclination may be of direct concern or it may be the only means by which a measure of displacement can be obtained.

The concern may be the lateral movement of the top of a relatively tall structure. In the absence of any other point of reference (e.g. an adjacent and stable structure) inclinometers may offer the only means by which this movement can be quantified. Their use would also be appropriate where there are movements of the facade of a building, or of a retaining wall.

Strings of inclinometers can be used to profile movements on both horizontal and vertical axes. At its simplest this is done by putting a line of sensors at reasonably close centres up a facade, across a floor, or along a beam. The output from each is monitored and the apparent movement calculated (making the assumption that the inclinometer is measuring the inclination of the chord of the section of the facade/floor/beam concerned, which is reasonable so long as there are no discontinuities in stiffness). To achieve higher levels of accuracy the inclinometers can be mounted onto relatively long stiff supporting members, which are fastened so that they are held precisely parallel to the structure at all times. When a line of these is interconnected so that the displacement can be accumulated over a distance, very high precision can be obtained. Precision of the order of plus or minus one tenth of a millimetre has been reported over runs of many tens of metres. An isothermal environment is generally a prerequisite for this order of accuracy. It has to

Figure 5.15 The tilt monitoring unit provides accurate information about changes in inclination. It can be demounted and remounted without loss of alignment.

be said that it is a rather intrusive technique which can only really be applied where appearance is of no concern or where the equipment can be hidden away.

Inclinometers suitable for use in monitoring building structures use one of four different technologies though all involve monitoring the effect of gravity on a pendulous mass. The pendulous mass may be a solid object or a small volume of fluid.

Probably the most readily intelligible forms of inclinometer use non-contacting position or displacement sensors to monitor the change in the relative position of a pendulum. The accuracy of such devices depends on the performance of the hinge mechanism, the position sensor and the length of the pendulum. The design of these devices always has to overcome the fact that the restoring force diminishes as the pendulum approaches the vertical null point. They are prone to hysteresis due to friction in the hinge mechanism and are liable to be disturbed by vibration and swaying unless damped (which usually compounds the hysteresis problem). They can also be affected by a high level of cross-axis sensitivity. The output from these devices will depend on the type of position sensors and associated signal conditioning employed. It is very often high level, either a DC voltage or current.

Accelerometers have also been used for monitoring changes in inclination. The simplest form of accelerometer consists of a mass supported and enclosed within a housing in such a way that the change in load in the structure supporting the mass can be monitored. Conventional strain gauges are frequently used though these are now being replaced by piezo-resistive devices. The output will depend on what signal conditioning is incorporated.

More sophisticated accelerometers sense actively, rather than passively, the effect of gravity on a pendulous mass. These use the combination of a torque motor and a position sensor to hold a pendulous mass in a fixed position with respect to the case of the accelerometer. The current supplied to the torque motor is a precise analogue of the inclination. These devices can offer very great precision indeed, and also offer a very high frequency response (which can be vital). They provide relatively low cross-axis sensitivity and are relatively insensitive to motion in the horizontal plane. Unfortunately they are exceedingly expensive.

The fourth group of inclinometers use fluid as the gravity sensing element. One version consists of a phial of fluid which wets conductors according to the attitude of the phial. This alters the resistance between the conductors and provides an analogue of the change in inclination. This device provides two arms of a bridge circuit and with appropriate signal conditioning such devices can offer very high and enduring precision over the limited range of movement they are designed to accommodate. Another version uses the fluid to change the capacitance of complementary pairs of capacitors and with signal conditioning provides a high level output. Again, high and enduring precision can be obtained.

5.3.6 Height and level changes

Changes in height or level are frequently of interest when a structure requires close monitoring. Reference has already been made to how changes in level can be ascertained using strings of inclinometers. There are several other techniques which can be applied in appropriate circumstances.

Measuring and monitoring level changes require a reference level to be transferred from a reliable datum point to the point of interest so that the difference between the levels may be measured directly and accurately. Various ways of achieving this in automated systems have now been developed.

Traditional surveying uses an optical approach and this has been used also in automated systems. Rotating lasers, commonly used for setting-out purposes, can be used with special detectors to discern changes in level automatically. The detectors consist of a vertical array of light sensors combined with electronics, which determines the average of the position of the upper and lower boundary of the portion illuminated at each sweep of the laser. The resolution of these devices is typically quite coarse (2–5 mm) and the accuracy depends entirely on the performance of the rotating laser. Apart from the performance of the detectors, the phenomena affecting accuracy are exactly the same as those affecting the accuracy of conventional optical levelling systems.

An alternative approach uses static lasers projected at a photo-detector which simply determines that laser light has or has not reached it. At one or more intermediate points along the light path devices are placed. These have blades which are in turn automatically lifted, lowered or rotated, with the intention of temporarily interrupting the passage of light. The interruption

Figure 5.16 A level sensing station undergoing trials.

of light is detected and knowledge of the precise position of the blades at that time allows changes in level and alignment to be determined.

A passive variant of the same technique uses a series of blades with accurately aligned holes to allow the passage of light until one or more blades moves out of alignment. When this happens the blade or blades which are responsible for cutting off the light can be identified and, if required, repositioned. The extent of the movement which has caused the passage of light to be interrupted can be determined from the amount the blades have to be moved to allow light to be transmitted. This, however, amounts to a level change detector rather than a measurement system.

Each of these techniques is capable of application in a variety of circumstances. All of them require a line-of-sight between the datum position and the point of interest. Where this is not available multiple lasers may be required.

Level information can also be passed from the datum point to the point of interest using a liquid reference system. This circumvents the necessity for lines-of-sight.

One of the simplest techniques for establishing the relative level of two points is, of course, the manometer or U-tube (see also page 15). The relative height of the liquid in one arm of the tube is subtracted from the relative height of the meniscus in the other arm of the tube to determine the relative level. This technique has been successfully automated by sensing the height of floats at a number of points around a building. In these systems float chambers are each interconnected by a system of pipes so that a liquid may flow freely between them. At each float chamber, the height of the float relative to the case of the float chamber is monitored and reported to remote data-acquisition equipment. The relative change in level is then calculated with respect to a datum float chamber.

The performance of such systems depends on the ability of the liquid to move freely between the float chambers and therefore relatively large-bore piping is required. Surface tension and surface wetting have to be taken into account in the design of the float and float chambers. Temperature will also affect the accuracy of the system if it is not installed within a near-isothermal environment. With appropriate care submillimetric precision can be obtained.

The liquid and float chamber approach obviously requires that all the float chambers are mounted close to the same level, although this may not be readily achieved. The requirement that liquid flows to each float chamber as the chamber changes level also limits the responsiveness of the system to rapid change. Greater freedom in the positioning of the sensing element and the character of frequency response that is required when monitoring the response of a structural element to live loads, can be achieved with a closed fluid system. Such a system has been designed for use in a specific heavy civil engineering construction task and this system has now undergone further development into a modular automated level sensing system capable of application in a wide variety of structures.

The principle is straightforward. The pressure in a liquid is proportional to the depth: measure the pressure at two points in a system of liquid filled pipes and with knowledge of the liquid density it is possible to determine their relative level. Achieving enduring accuracy is rather more complex. This approach has been used successfully to achieve millimetric accuracy on a cable-stayed bridge of 275 m span in the context of large movements resulting from combined traffic and wind loadings.

5.3.7 Behaviour of building fabric

So far the sensing technologies described have been concerned largely with determining the relative position and changes in the relative position of elements of a building's structure. Automated monitoring can also be applied to measurement of the behaviour of the materials which it comprises, in particular to measurement of local changes in strain. This will normally entail bonding or embedding strain gauges into the material itself, or fastening an external strain transducer onto the fabric.

Foil-style strain gauges generally consist of a conductor laminated to a carrier in such a way that stretching or compressing the carrier (gauge) along its sensitive axis or axes will similarly stretch or compress the conductor. The conductor can be a filament arranged in a zig-zag pattern but more usually it consists of a foil which is etched or cut to form a zig-zag pattern. A change in resistance results from stretching or compressing the conductor which then provides an analogue of the change in strain.

Foil strain gauges are used in Wheatstone bridge circuits: quarter, half or full (or simulated) bridges can be employed. Temperature correction must be applied if accuracy is to be obtained. The thermal characteristics of the gauge are usually matched with those of the material to which they are applied, but nonetheless temperature compensation is still required because the resistance of the conductor changes with temperature as well as with changes in strain. Half and full bridges can be arranged to be self-compensating, providing that gauges with identical thermal characteristics are used. Some data acquisition systems provide facilities for using a single unstressed gauge as a temperature compensation gauge supporting a network of strain gauges.

Foil strain gauges are normally bonded to the fabric at the point or points of interest using special adhesives. Weldable gauges are also available for use on structural steel. The bonding system and the technique of bonding in both cases is crucial to the performance, durability and endurance of the measurement system. Installation has to be carried out with great care by suitably skilled personnel otherwise creep or delamination will destroy the value of the data.

Vibrating wire strain gauges use the principle that the natural frequency of a tensioned element varies with the tension applied and they measure the tension in a wire changed by relative displacement of the ends of the gauge. When a measurement is required the tensioned element is plucked electrically and the natural frequency is measured. Vibrating wire strain gauges are generally much larger than foil strain gauges. Their size prevents them from being used for very localised measurements of strain; in effect they are displacement transducers with very limited ranges. They are normally mounted using brackets which are fastened to the material which is being monitored although the nature of the gauge and the mounting arrangements can result in temperature effects masking the data: temperature must also be monitored so that corrections can be made.

Foil strain gauges are available which have a baseline for measurement between one and three hundred millimetres. Vibrating wire gauges necessarily span several centimetres. Over the longer baselines it becomes practicable to measure strain using extended displacement transducers. A device with a range of only one millimetre can be mounted on an extension piece which extends its baseline of measurement to around 0.5 m. Provided accurate temperature compensation is carried out, microstrain resolution can be achieved. Extending the baseline overcomes local variability in the material monitored and reduces the disruptive influence that the mounting and fastening arrangements have on data quality.

5.3.8 Load

Strain gauges can be used to infer values for load changes in structural elements. Where the structural element is relatively simple in nature, this can be achieved with a reasonable degree of accuracy and assurance. In more complex situations, load cells can be introduced into the structure. These are in essence calibrated structural elements which may be used to monitor tensile and compressive loads.

The scope for application in prolonged monitoring of existing structures is limited unless the structure is modified to allow load cells to be incorporated. To date this has only been done in exceptional circumstances. Columns have been cut and jacks and load cells used to measure the force required to transfer the load in the column to the jacks. For longer term monitoring a portion of the column can be replaced by one or a number of load cells. Similar procedures can be used to measure and monitor loads in tensioned elements.

Load cells generally consist of a load sensing element to which strain gauges are fixed. The load sensing element will be designed to carry the designed safe working load and to deform in a predictable linear, elastic fashion up to and indeed far beyond this load. The load sensing elements in both compression and tension load cells are normally annular in form. Many tension cells are simply compression cells with the load applied through the annulus.

An alternative form of load cell is the single and double shear beam load cell. Single shear load cells provide a means by which loads can be measured eccentrically. Double shear load beams measure the load on the central portion as it is transferred to the flanking portions. They are typically employed in linkages where lateral restraint is required.

Load cells can be constructed to measure loads ranging from a fraction of a gram to thousands of tonnes. The most common operate in the range between 50 kg and 50 tonnes. Outside this range special design is required.

Load cells typically offer infinite resolution. Very high accuracy (better than 0.5% of the range) can be obtained by meticulous calibration and correction. More modest accuracy (better than 2%) is readily achieved. Most load cells will have reasonably good thermal characteristics with a measure of built-in temperature compensation.

5.3.9 Environmental parameters: temperature, wind, humidity and moisture content

Automated monitoring systems are frequently capable of discerning the effects of environmental factors such as ambient temperatures, solar heat gain, and geotechnical factors including changing groundwater levels. Obtaining measurement of these factors will allow correlations to be made and the underlying trend of movement to be elucidated rapidly. When a trend is concealed by daily and seasonal weather changes, the correlation between a crack width and fabric temperature can only be demonstrated if all appropriate information is available.

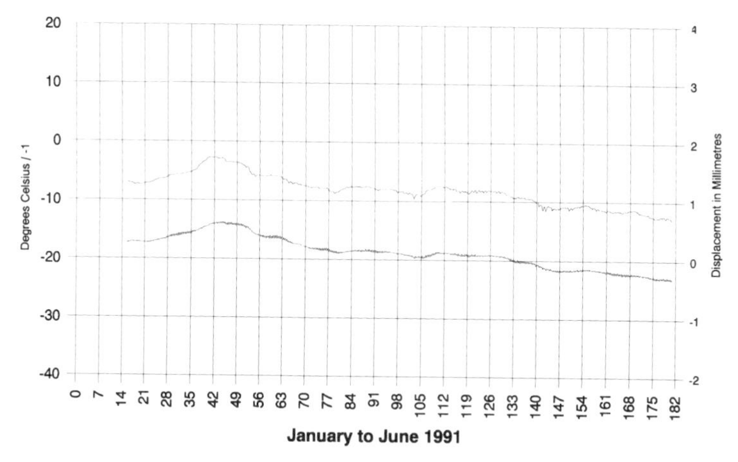

Figure 5.17 Simultaneous measurements of movement and temperature obtained at hourly intervals over a six month period. The upper trace is the amount of movement (in mm). Note the close correlation between the shapes of the movement and temperature traces with a slight time lag in the response of movement to temperature variations.

Temperature monitoring is a very common requirement. Consequently there is a wide range of devices available which covers the measurement of temperature in almost any setting and to extreme temperature limits.

Frequently the two most important factors determining the thermal behaviour of a building will be the external and internal ambient temperatures. In buildings the internal ambient temperature will normally be controlled so that the internal fabric (including structural elements) will not vary greatly. The external fabric and outer structural elements on the other hand will vary considerably with ambient external conditions.

In damp and cloudy conditions the behaviour will be reasonably uniform. In hot sunny conditions quite abrupt changes can occur due to variations in light and shade. Unprotected structural steel will heat up at very high rates when exposed to strong sunlight.

Just as the fabric of a building will be warmed by sunlight, the material of the sensor will be heated if it is not shrouded in a suitable manner. Meteorological housings can be obtained for temperature sensors and these have light reflective and insulation properties which conform to standards laid down for air temperature monitoring. Alternatively, the temperature sensor can be sited to obtain a measurement which is indicative of the conditions pertaining to the part of the structure which is being monitored or to part of the monitoring system itself.

Just as sunlight falling on the sensor can affect the readings obtained, a variety of other factors can play a part. Draughts of cooler or damp air can depress readings by several degrees. A large body of water will hold its temperature through the night and evaporation will lower the air temperature above the water surface. Hot air from an extractor fan can cause considerable problems. Siting temperature sensors therefore requires some thought if a measurement is to be obtained which is to be of value. The use of long tensioned wires in measurement systems requires accurate temperature measurement if temperature correction is to be achieved with any precision.

Measuring fabric temperatures requires even more careful consideration. For a start, many temperature sensing technologies have a self-heating effect although in normal operation this is not a problem. The heat generated by the current which is passed through the device in order to determine the voltage drop (and in turn the resistance and the temperature) is quickly dissipated to the surrounding atmosphere and the resulting temperature change of the temperature sensor is either negligibly small or predictable. Insulate the sensor and the error can be greatly magnified.

Temperature sensors are frequently attached to the associated data-acquisition equipment using copper conductors. Copper provides a very effective transmission medium for the signal but it also provides an effective means of transforming the heat picked up in a surface mounted cable from solar heating to the sensor. The very act of drilling a hole to receive a temperature sensor may have an effect on the temperature which is to be measured. Surface temperature measurement also requires careful consideration, particularly if sunlight falls on the surface.

It is fortunate that a requirement for great precision in fabric temperature measurement is fairly rare. All that is sought in the majority of cases is a body of data which provides a guide as to how the structure responds to heating and cooling.

Monitoring wind speed and direction is rarely called for in connection with building structures and occurs more frequently in major civil engineering works. However, it can be relevant where the scale of the building is very large or where the incident winds are unusually strong.

Anemometers and wind direction indicators which provide an analogue output can be coupled readily to automated structure monitoring systems. To do so in a meaningful fashion requires careful consideration regarding the monitoring strategy as wind speed and direction vary continuously. If the data is recorded in a continuous fashion then an entirely respectable analogue will be obtained. However, if the recording consists of discrete readings taken at regular intervals, instantaneous values may well be meaningless. More useful data will be obtained if additional signal conditioning and/or data-processing is incorporated. The inputs can be filtered so that the values obtained for wind speed and direction are averages for the period between readings. Equally the maximum and minimum wind speed can be recorded but the system will require either hardware or software to reset itself after each record is taken. Erroneous data can be recorded, for instance if the wind veers more than 180 degrees the value recorded could show that it has backed, unless the recording interval is short enough to reveal the true nature of the change.

Humidity is easily monitored and various sensors are available from the fields of process and environmental control. Sensors used in heat, ventilation and air conditioning systems are frequently appropriate as these offer suitable ranges and reasonable accuracy. Process control sensors can be selected when the humidity is likely to go outside the range which is normally met within buildings intended for human habitation.

Humidity can be measured in terms of percentage relative humidity and/or dew point. The former indicates how close or far the atmosphere is from saturation and the latter indicates the temperature at which condensation would occur given the current atmospheric moisture content. The two values are related for any particular gas or atmosphere.

The majority of humidity sensors use the principle that changes in the moisture content of the dielectric between the plates of a capacitor will alter its capacitance. Special dielectrics are used which will take up or release moisture depending on its availability in the surrounding atmosphere. Circuitry is then used to detect the consequent changes in capacitance and to provide a signal. In the past such sensors have been prone to problems with saturation and pollution but now some sensors incorporate means by which the sensor can purge itself or be purged. Even more recent developments have seen these problems and requirements overcome and devices are available which offer enduring accuracy over a limited range (<0 to 100% relative humidity) without the need for periodic maintenance and recalibration.

An alternative technique is the direct measurement of the dew point. This is achieved by controlling and measuring the temperature of a mirror and using an optical detector to observe when condensation forms on the mirror surface. This technique offers greater precision but at significantly greater cost.

There is sometimes concern about the changes of moisture content in building materials. The interest may arise from the properties of the material itself or from a need to check that remedial measures to prevent further ingress of water have been effective.

Unfortunately monitoring changes in moisture content of solid materials on a continuous basis is usually difficult to accomplish with meaningful accuracy. The best traditional technique involves removing samples and weighing them before and after de-watering. This is not a process which can be automated easily, and certainly not *in situ*.

Hand-held moisture meters work on the principle of measuring the resistance between two pins pressed into the moisture bearing material. This provides a reasonable means of ascertaining the moisture content of a wall or slab, although it is prone to inaccuracy. Salts dissolved in the water will seriously affect the perceived moisture content and the presence and behaviour of hygroscopic salts will affect readings depending on the amount of water they take in or give up to the atmosphere. Nonetheless in ideal circumstances, or where little precision is required, this technique can be applied when using automated monitoring techniques. It is crucial that any conclusions take full account of the context within which the measurement system operates.

Another approach is the measurement of the equilibrium relative humidity in a void created within the material concerned, or in an insulated box fastened with a reliable seal to the surface of the material. Again this approach has to be applied with discretion and it will only be effective while the atmosphere within the void or enclosure is not saturated. It can therefore only be used with relatively dry building materials and the installation must itself not affect the moisture content of the surrounding materials. A sheet of polythene laid over a dry concrete slab will result in condensation on the underside of the polythene and a large wet patch forming on the floor. Equally a plate mounted over an opening in a wall containing a humidity sensor will cut off the route by which moisture would have evaporated and will affect the local moisture content of the wall accordingly.

5.3.10 Geotechnical parameters

Groundwater levels and changes of groundwater level can be monitored relatively simply using a variety of techniques normally involving drilling a borehole. When this is done careful consideration has to be given to how the work is carried out and to how the borehole is to be lined. In many instances these will both affect the reading which is obtained. The possibilities include schemes in which either the borehole only fills and drains through the bottom, or it fills and drains through openings in the casing at a particular height, or it is free to fill and drain at any height. Whichever strategy is adopted, the measurements will be affected if rainwater is allowed to enter the borehole from the surface.

Liquid levels can be monitored using a float in a tube, on a guide or located by a swinging arm. Providing that the float does not sink or the mechanism does not jam, the float will track the level of the liquid surface. The position of the float may then be measured in one of a number of ways. A displacement sensor can be used to measure the change in height of the float directly or to detect the change in angle of the swinging arm which locates the float. More sophisticated inductive measurement systems measure changes in the position of the float as it moves along a sensing rod which doubles as a guide rod. Each measurement technique has its own blend of merits and demerits, and they all have quite limited ranges. Their usefulness will be determined by how easily they can be installed or applied to the task in question.

An alternative approach is to measure changes in the head of liquid over an immersed liquid pressure sensor. Pressure sensors are made which are specifically intended for this purpose, and they incorporate an atmospheric balance tube in the signal cable so that changes in atmospheric pressure do not affect the reading obtained. These sensors are lowered into the water until they are below the lowest anticipated water height and are then tethered. Subsequent measurements all refer to the starting point so any subsequent movement of the sensor will introduce an error unless the sensor can be repositioned accurately. The output from these sensors is frequently a high level one (a voltage or a current) which very often will be directly compatible with the data-acquisition equipment which is employed. In some cases the output is low level which may require additional signal conditioning.

Water levels in rivers and reservoirs are now frequently measured using echo-sounders encased in relatively large diameter tubes. The echo-sounder is mounted above the fluid surface and bounces a signal off the surface of the water to determine the relative position of the water surface. The great benefit is that it is a non-contacting system. Apart from the tube or casing there are no immersed parts to the system. However, accuracy and resolution are not particularly good. With better systems these are expressed in terms of a small number of percent of the measurement range. Ranges can be as small as a few tens of centimetres up to many metres. The size of the casing required increases with range as the primary echo from the target water surface area must at all times remain strong in relation to the background noise, including the signal emitted by the sounder which takes an indirect route to and from the fluid surface. There is an obvious advantage to be gained from using devices with smaller ranges wherever practical.

The majority of these techniques for water level measurement have been used down boreholes in the past. The first group of mechanical devices presents installation difficulties. The pressure measurement and echo-sounder techniques are relatively easy to use.

Pore pressures may also be monitored automatically and autonomously. The measurement technologies are relatively straightforward but their application is not. The problems are concerned principally with the siting of the transducers and their installation.

Other monitoring of a geotechnical nature which can readily be undertaken automatically includes tasks like monitoring the stability of embankments and retaining walls using inclinometers, settlement profiles using inclinometer strings, settlement using buried datums, rods, wires and displacement transducers, and lateral ground movements using buried datums, rods, wires and displacement transducers. Installation and environmental considerations pose a slightly different set of problems.

5.4 Recommendations

This chapter has given the reader a superficial overview of what is possible when automated structure monitoring is considered. The purpose has been to provide readers with a feel for the subject so that they can confidently obtain systems, equipment and supporting services appropriate to their needs enabling the effective management and execution of the processes of determining and defining requirements and seeing that those requirements are then satisfied.

The remaining paragraphs give a rudimentary list of recommendations and questions relating to automated and autonomous monitoring:

- Define the problem to be tackled. Define the purpose or purposes of the system clearly. A concise narrative statement of the purpose is generally much more useful than a list of measurements which are to be made.
- Define the requirements in terms of the problem, not in terms of the instrumentation.
- Define which parameters are to be measured. Do not dictate how they are to be measured unless there is an overriding reason to do so.
- Define the range over which each parameter is to be measured. If different, define the range over which each parameter may vary. Avoid quoting ranges which are large enough to cover all eventualities when concerned with behaviour within a much smaller range.
- Define the accuracy which is actually required, not how it is to be achieved. If great accuracy is only required over part of the range make this clear. Achieving high accuracy over extended ranges will be very much more expensive than achieving it over a limited part of, or at one point in, the range.

- State the size of the smallest change in a parameter which is to be discerned, in numerical output and in graphical output.
- Is absolute accuracy or only incremental accuracy required? It is generally much more demanding to ask, for example, how far one point is from another than to establish how far it has moved with respect to another.
- Determine or estimate if possible the minimum length of time it will take for a change, equal to the smallest change required to be observed, to take place. Do the same or characterise the period required for a change 10, 100 (and, if appropriate, 1000) times as large.
- Decide what sort of output is required and the reasons for wanting it. At what interval are fresh data required? Consider carefully whether it will be possible to make use of data obtained at such short intervals given the form of output proposed. Equally, is the interval too long? What are the implications of not reviewing the structure's behaviour at close enough intervals?
- Are there circumstances in which it will be appropriate to decrease or increase the amount of data obtained? What are the implications for data-handling, processing and presentation of increased volumes of data?
- How, when, and where is the data to be handled? And by whom?
- What will handling the data involve? If things are as expected? If not?
- Does the system which is to be installed need to draw attention to the fact that something unusual has happened when it happens? What are the criteria? What information will be required when this happens? Where?
- Give some thought to the future. What is the required service life of the system? What is permissible/possible in the way of maintenance access?

Automated structure monitoring has enormous potential, if applied sensibly. Above all it has the capacity to release resources, which would otherwise be spent in obtaining data, to be applied to the task of determining what is or is not going on.

6 Planning a scheme

G.S.T. ARMER

6.1 Introduction

Before any scheme for construction monitoring is developed, the requirements of the instigator of the plan should be clearly articulated. These requirements obviously have to be tempered by what can be achieved with any practical monitoring system. It is especially important that the proposer of a scheme should discuss with his client, who will normally be either the building owner or the occupier, the benefits of monitoring *and* its limitations. It would be a disservice to claim for a scheme the function of a unique panacea for any real or even presumed set of structural problems.

It is a fact of life that the solutions to virtually all engineering problems are dependent upon the monies available to achieve the desired end. Construction monitoring is no exception. Consider the matter firstly from the point of view of the property owner. The amount of money available for monitoring a particular property will depend, *inter alia*, upon the value of the property itself and on whether it is part of a population of similar buildings in single ownership. The financial implication of shortcomings discovered in a single property may be considerable if replicated in a number of other properties. In such a circumstance, it may be appropriate to limit the effort spent on each individual structure and to broaden the scheme to cover the whole population.

An example of this approach is the concern with safety which resulted in investigations of non-traditional housing by the Building Research Establishment (BRE, 1987; 1988; 1989). These were followed by nationwide (in the UK) programmes of survey and refurbishment of the estate of low-rise prefabricated concrete houses, of which there were around 180 000 units.

The amount of money worth investing in a scheme may also depend on the letting or selling potential. Interestingly enough, through one of the many paradoxes that we have to deal with in the world, it may be that a strong sellers' market and a very weak/non-existent market such as might occur in an economic depression could both lead to a decision not to invest in a property. For on the one hand there is no need and on the other—no point!

Secondly, the occupier of a building may see the situation with quite different eyes. The hazard to business prospects and the safety of staff may well colour the view taken of any scheme. The cost of establishing a comprehensive monitoring system could well be justified if it avoided the difficulties which might be encountered as a result of a failure of the occupied premises. Alternatively, if a system is installed, there may be problems with staff, who may wonder whether or not it is safe for them to stay at their places if they see wires and gauges fixed all over the building.

There are then decisions which have to be made by the client (in consultation with the engineer) which do not have an engineering base but which will have a profound effect upon the form and magnitude of any scheme for monitoring.

6.2 Schemes for populations of similar structures

What constitutes 'similarity' between two structures can vary widely according to the particular circumstances. For example, common materials of construction, e.g. brick masonry, timber, etc., common elements such as prefabricated steel frames or trussed rafter roofs, common functions such as a swimming pool, all provide a link sufficient to identify a population and to establish the membership or otherwise of a particular building. This list of characteristics is by no means exhaustive. So it is clear that a population of similar structures is comprised of a group of buildings which have at least one common feature. If some shortcoming of such a common feature gives rise to the possibility of investing in a monitoring scheme in one property, then because it exists in other buildings which *ipso facto* become suspect, the population is then, in this circumstance, of significance.

Monitoring, by definition, is concerned with the identification of changing response or condition within a structure. Such changes are measured against a time-scale, offering the option of choosing the frequency of observation from continuous to once in the lifetime of a structure. For the situation where there is a population of structures, the problem of what to survey, where and how often is exacerbated. In particular, how can measurements taken over a period of time on one building be equated with a similar number of equivalent measurements taken quickly on different buildings within an identified population? Unfortunately, for all practical purposes, one is reduced to using heuristic arguments which have only limited support at the moment. However, there are a few recent examples of population studies which may be used for guidance.

In 1973, two prestressed concrete beams made with high-alumina cement (HAC) and which formed part of the roof structure over a school hall collapsed and were investigated by BRE (1973). This failure was followed by another in a roof structure over a school swimming pool described by Bate (1974). The critical element common to these failures was identified as the conversion of the high-alumina cement concrete. In this context, conversion means the change in the structure of hydrated cement from a metastable to a stable form, which usually results in a reduction in the strength of concrete made with this particular type of cement. Since these failures occurred in public buildings and the hazard to life in similar buildings was clearly a possibility, the extent of use of high-alumina cement in structures was investigated. The result of this investigation was that a population of between 30 000 and 50 000 buildings constructed in the post-war period was found to have used this material. The manufacturers claimed that some 17 million square metres of flooring were constructed with this material. Following the appreciation of the scale of the potential problem, a series of activities was initiated and reported by Bate (1984). Firstly, a programme of experimental work was established to gain a better understanding of the phenomenon of conversion in HAC concrete, in particular the temperature and moisture conditions which affect the rate at which it occurs. These data were matched with samples of concrete from buildings within the population. Secondly, the effect on precast beams of strength loss due to conversion was the subject of an experimental project. Thirdly, the accuracy of non-destructive test methods for site inspections was investigated. Fourthly, there was also a limited amount of load testing of complete floors to establish some criteria for owners who felt it appropriate to adopt such an expensive option. These preparatory activities established a base for the owners of the properties at risk to survey the buildings in their care and to identify those structural elements in need of immediate replacement, those in which some conversion had taken place and which would call for remedial action before the useful life of the building had been consumed, and those elements which were in environments which did not support rapid conversion and were therefore not a risk to safety. Of these three categories, only the second one leads to a requirement for regular monitoring. The frequency of inspections for *every* structure in this more limited population should be determined on the basis of the thermal and moisture conditions prevailing in the structure and the existing degree of conversion and the known rates of conversion in particular environments.

In the early 1980s, quite serious problems were identified in a prefabricated concrete house construction system known as Airey. This was followed quickly by the discovery of structural problems in a wide variety of other non-traditional precast concrete housing systems which were tried in the post-war period (although a few predated this time), mostly for public housing schemes. The number of units involved was in the hundreds of thousands rather than the tens of the example discussed above. In the beginning, the major problem facing owners of housing stock was one of identification. Many of the systems were designed to have a similar appearance to traditional construction. This problem was even worse in Scotland, because there rendering external walls is almost universally adopted for low-rise housing. There were probably some 30 or 40 systems tried in one part of the country or another. Fortunately, it seems that only 17 systems were used for 98% of the stock. The problems were essentially related to the use of materials. High-alumina cement was one cause of concern in Orlit houses. The effect of chloride admixtures, described by BRE (1985a; 1986), used in Portland cement concrete to accelerate hardening and the carbonation of structural concrete discussed by BRE (1981) and Currie (1986) and the consequent reduction in alkalinity of the environment of the reinforcing steel was the basic factor suspected in all the systems.

The programme developed by the now defunct Scottish Special Housing Association (SSHA) for their estate of Orlit houses is particularly interesting for the present purpose of illustrating schemes for population monitoring. (The SSHA has now merged with the Housing Corporation in Scotland to become Scottish Homes, which is a government agency.) The initial population of Orlit houses owned by the SSHA numbered 1116, and they were distributed on eight sites throughout Scotland, as described by Mitchell (1989). The structure of the Orlit house most commonly comprises double portal main frames with secondary beams spanning between the frames, as illustrated in Figure 6.1, to produce a traditionally styled two-storeyed dwelling. The roofing system is either flat, comprising precast concrete planks with a waterproof covering, or the conventional pitched timber construction with a slate or tile finish. The first type inevitably lead to maintenance difficulties. There are a number of variations on this form but not affecting enough properties to be very significant in the context of the subject of this section.

The problems associated with the high-alumina cement concrete used in the stitch joints in the main frames were essentially simple although not cheap to resolve, since once they had been located the best solution was usually to replace the joint concrete. The problems resulting from corrosion of the reinforcement in the secondary beams were not so straightforward to deal with. The corrosion protection of the steel in the secondary beams proved to be highly variable depending on the chloride content of the concrete, the depth of cover, the depth of carbonation and the overall condition of the property. The result of having such a variety of potential causes of corrosion was that the visual evidence of problems ranged from conditions that reflected major degradation to those which were apparently perfect.

After a programme of inspection and general assessment, 326 units were demolished. A prognosis for the future life of the remaining units was established. The performance of the secondary beams was then the subject of a regular monitoring survey using optical probes sighting through holes drilled in the first floor and upstairs (in the flat-roofed units) ceilings. The holes were plugged between surveys so that they were not too unsightly for the tenants. The frequency of survey was *increased* after experience was gained in the first round, and it is now undertaken on an annual cycle. The inspectors were given detailed instruction on the interpretation of what they saw through the instrument so that consistent action could be taken in response to the reports.

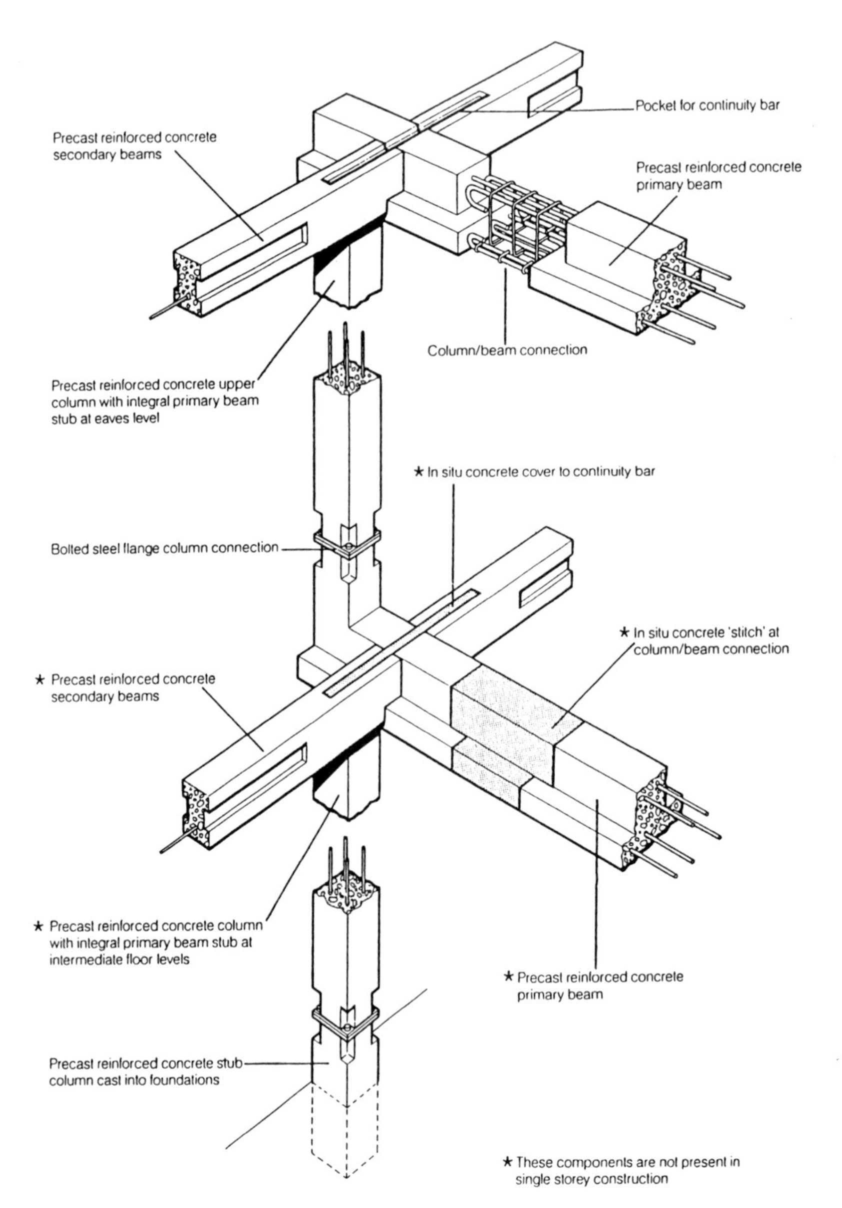

Figure 6.1 The structure of an Orlit house.

The programme of monitoring was therefore established after:

(a) A full-scale assessment of the population.
(b) The selection of those properties to be retained in the estate, on economic and management grounds.
(c) Identification of the critical features to be monitored, i.e. cracking due to steel corrosion in the secondary beams.

Another problem which arose in the early 1980s related to public housing concerned the safety of large-panel system buildings constructed in the 1960s. The first sign of potential hazard was a storey-height gable end cladding panel slipping because its fixing was inadequate. This was followed by a tenant finding a gap between his floor and the wall of his flat. This latter discovery was in the notorious Ronan Point, the subject of an inquiry by Griffiths (1968) and was sufficient to generate much public concern. As a consequence of the occurrence of these events, a nationwide survey of large-panel system-built flats was instigated, firstly concentrating on Bison and then Taylor Woodrow–Anglian buildings, following BRE (1985b), and then including all the other systems.

The format of the housing units provided by these systems ranged from simple low-rise to over 20-storey high-rise blocks. There was an equally wide-ranging span of problems encountered in the initial surveys. On the safety side, there was a distinction made between blocks of five storeys and over, which had to be designed explicitly for accidental loading such as a gas explosion, and those blocks under five storeys, which did not have to meet this requirement. The recommendations for appraisal were therefore formulated to match this distinction.

The objective of appraisal of the system-built housing to assess its safety was encapsulated in the following three questions:

(a) Is the building safe now and will it remain so during its service life under normal and accidental loads?
(b) Is there now or likely to be during its service life a hazard from falling debris?
(c) What are the maintenance and servicing burdens on the owner likely to be?

The following monitoring programme was then developed:

All large-panel system buildings required to exceed 25 years' service life from the date of construction should be subject to full appraisal for structural safety and durability.

(1) For buildings of five or more storeys, the appraisal of safety should be made with respect to both normal and accidental loads.
(2) For buildings of less than five storeys, the appraisal of safety should be made with respect to normal loads.

Subsequently for all such buildings:

- Visual inspections of the external envelope, including parapets, balconies, etc., to identify potential hazards from falling debris, should be made at intervals of 1, 2 and 5 years following the initial appraisal and then at minimum intervals of five years.
- Visual inspections and assessment of *in-situ* structural joints likely to be accessible to rain penetration of the outer envelope, at 10-yearly intervals.
- Full appraisal of the complete building at intervals of 20 years.

Because of the short period during which this particular population of buildings was constructed, for those buildings not required to remain in service for longer than 25 years from the date of their construction it was considered sufficient to undertake only a visual survey for falling debris hazards and to check concrete chloride content.

With such a large population of buildings some difficult decisions had to be taken regarding priorities. The following approach seemed to be the most appropriate.

Buildings should be ordered according to:

(1) Previous in-service maintenance history.
(2) Age.
(3) Number of storeys.

The foregoing examples of population monitoring illustrate the complexity of the decisions to be made when embarking on this course of safety maintenance. It is reasonable to expect that any scheme which is established for a particular class of problem will necessarily be subject to modification during its operation. Indeed, it would be foolhardy and probably an economic mistake not to allow for such development.

There are other sectors of the construction market in which monitoring schemes are commonplace. These sectors are transport, notably bridges, discussed by Sowden (1985), the offshore oil industry and the critical structures for nuclear power generation. It is difficult to establish the effectiveness and cost benefits of these schemes since they are associated with structures subject to much greater quality control during their construction. Their installation is fundamentally a reflection of the limited knowledge of either how the particular structure will behave in service or the loads and deformations to which it will be subject, or a combination of both of these factors. One long-established monitoring scheme for buildings described by Ishizuki (1981) was started in Japan in 1969 and involves inspections of 4000 government buildings at 5-yearly intervals. The objectives are to establish practical service lives for building materials and components in a variety of working conditions. In this survey, the components are classified by, *inter alia*, a degradation index. The cycle time seems rather long for some components, especially those which are approaching the end of their useful life. The value of such schemes will be principally as generators of data banks which can be quizzed to aid the interpretation of single condition surveys.

6.3 Schemes for important buildings with degrading materials

Most degradation of structural materials in buildings reflects the progress of chemical actions. In timber the principal agents for these actions are, of course, biological. The degree of degradation is dependent upon the stability of the particular chemicals present, the temperature and the presence of water in some form or other. The offending chemicals have sometimes been incorporated in the material and sometimes are in the atmosphere. Sometimes they are not 'offending' in the conventional sense but have been used in unsatisfactory combinations which, for example, facilitate electrochemical reactions, as in the case of some mixed metal designs, or in the case of timber are generated biologically.

It is clear that for important historical buildings schemes for regular inspection as described by Chambers (1976) are invariably associated with regular maintenance programmes. Feilden (1985) characterises the relationship between the two activities thus:

- Regular inspections (of buildings) enable architects to develop a strategy for preventive maintenance.
- The architect will gain valuable feedback on the performance of various materials and learn how his buildings are used and abused.

This combined activity is expensive, and for many historic buildings impractical for financial reasons. One particular group of such buildings comprises the Church of England churches. It is recognised by the Church authorities that even structural surveys at 5-year intervals are beyond the purse strings of most parishes. The Church of England (1980) have therefore offered the following model brief:

> The inspection of the Church is to be visual and such as can be made from ground level, ladders and any readily accessible roofs, galleries or staging, and only selected areas are to be examined in detail; parts of the structure which are inaccessible, enclosed or covered, are not normally to be opened up unless specifically requested. Inspection is to include as far as practical all features of the building covering all aspects of conservation and repair . . .

Viewed by an engineering observer this might appear to be a horrific way to attend to the needs of these special buildings; it is, however, the only pragmatic solution to the problems of the enormous burden of the Church estate.

The signs of problems in old buildings are usually associated with movement of some sort. For example, the effects of moisture movement and temperature changes need to be identified and assessed for stability (even though they may be cyclic) and for their rates of change. There are now a number of gauges and techniques available to monitor such movements (Collacott, 1985) but the big problem, apart from the cost, is where to fix such instruments and how to interpret the results if changes are recorded. Mainstone (1988) has described the structure of one building, the Hagia Sophia in Istanbul, in great detail. This church's construction and reconstruction history is complex, requiring skill and time to make sense of measurements made in any monitoring scheme. Such a building would not be recommended for the beginner in these matters to try out his talents!

The CIRIA (1986) report on renovation recommends monitoring as an aid to the development of an understanding of the behaviour of a building structure. The report also offers the following comment:

> Most defective old buildings give plenty of warning signs before they become dangerous. In this respect, many buildings which may appear to be unsafe, according to a simplified calculation of the strengths of the individual elements, continue to resist load without apparent distress.

The degradation of masonry is often the direct result of pollution (see DoE, 1989), but the mechanisms of material breakdown are extremely complex and related to many factors in the environment. There is, however, a device called an 'immission rate monitoring apparatus' which measures pollutant absorption, and data from this type of equipment are quoted in $mg/m^2/day$. The damage to some types of stone has been measured at 33–55 μm per year, which is equivalent to immission rates up to 128 $mg/m^2/day$.

Monitoring rates of 6-monthly intervals would give useful data on progress towards structurally significant conditions. Continuing changes in old buildings related to thermal and moisture movements of masonry are likely to be relatively slow, and monitoring rates of less than once a year are unlikely to be necessary. The sites for movement gauges or reference points will often be identified by existing cracks. Deformations of load-bearing walls could be precursors to buckling instabilities, and if such mechanisms are a possibility more frequent readings would be appropriate.

Moisture and thermal changes have rather different implications for timber and metallic structural elements. Timber is sensitive to moisture for two reasons; the first is its mechanical properties, e.g. as demonstrated by Gosselin *et al.* (1987) and the second is its susceptibility to attack by a variety of organisms. The result of changes due to both these phenomena would be deformations of one kind or another, therefore the usual range of measuring instruments are appropriate for a monitoring scheme and should be used on a cycle time of not less than 3 months. Thermal changes are only likely to be significant indirectly by altering the moisture regime in the timber.

The corrosion of metallic components in building is a subject with a long history. Again, it is the combination of temperature, moisture and pollutants which generate corrosion and there is a considerable body of data on rates at which the basic material is degraded. It is unlikely that cycle rates need to be shorter than 1 year for this problem.

Thermal changes, aside from their effects on corrosion rates, can affect metallic elements principally by causing geometrical variations which may differ in size from elements of other materials in the structure. Such differential expansions or contractions would be exhibited by distortions of the structure and therefore could be monitored by the normal range of movement gauges. Cycle times should probably be short if problems of this type are suspected, and daily measurements will be necessary until a clear picture of any progressive behaviour is established.

6.4 Schemes for important buildings subjected to major disturbance, such as tunnelling, piling, etc.

Schemes of this type are undoubtedly the most widely used of all the possibilities discussed so far and therefore the most usual applications of the techniques described in the other chapters. From many points of view they are also the easiest to design. There are two reasons for this: one is that the potential cause is clearly identifiable, e.g. a tunnelling development, a new building constructed near the foundations, etc.; the second is that the duration of the disturbance will usually be known fairly accurately. If an important building is likely to be affected by nearby development, then it should be instrumented in such a way as to ensure that any significant distortions of its main load-bearing structure will be detected.

The instrumentation should be installed some weeks before the disturbing construction is started so that the normal movements of the building are recorded. The cycle time should be 4–6 hours so that the detail of the diurnal variations can be established. It is probable that for major works a cycle time for the system should be hourly during the time of major hazard. An important feature of these schemes is that monitoring should continue for at least 1 year after completion of the disturbing works and ideally longer, say for 2–3 years. The cycle time for this final stage of the programme would not normally need to be shorter than once a week.

There is a problem unique to cities with much high-rise construction and where wind is likely to be a major design load. Very tall buildings affect the character of the wind a long way from their own sites. It is possible then that the construction of a large building in the environs of an existing building could cause major changes in the wind loads on it. If this is at all a possibility then it would be sensible to install some accelerometers with a recording system, activated at a reasonably high wind speed, to monitor structural response. Ideally, some data on the behaviour in wind storms before the new construction would be of considerable help in the interpretation of the new information, but obviously the need for such measurement is difficult to predict.

6.5 Schemes for long-term mechanical damage, subsidence, sonic boom, traffic noise

Long-term mechanical damage in buildings will most often be caused by some sort of traffic movement. Floors will usually be protected by a wearing surface which is essentially non-structural and therefore will not need any monitoring system to detect potential structural hazard. However, the same may not be true for columns and walls. It would therefore be advisable to monitor the condition of these elements in factory and car-parking structures on a 3- to 6-monthly cycle. Subsidence and its concomitant in this context, heave, is usually a problem in well-identified areas such as those where mining has occurred or is still active, or where the soils are particularly affected by moisture movement. Measurements should be of structural distortions and cracking, and although it is difficult to recommend a cycle time which is both technically and economically appropriate something around 6 months seems a reasonable compromise. The final decision must depend on the local circumstance since, for example, if mining is proceeding immediately underneath the building then much more frequent measurements should be taken.

Problems due to traffic noise and sonic boom are quite difficult to resolve since the owners of the properties are usually much more distressed than their buildings and they often relate their own level of discomfort to that of their structures. It is vital that any scheme to monitor structural response to these environmental problems is established only after a detailed survey has identified every existing defect in the building. It is unlikely that cycle times need to be less than 3 months for this class of problem.

6.6 The 'intelligent building'—the future for monitoring

The developments in expert systems, instrumentation and economical data acquisition make the future for the in-service monitoring of building structures an exciting probability. Already there are some systems in operation. One of the oldest is part of an active damping facility in a North American high-rise building which is activated when the dynamic movement of the structure surpasses a preset limit. Another is a system for detecting subsidence in a mining area which is associated with permanently installed jacking to relevel the building after it has reached a predetermined distortion. Yet another is a very large roof structure over a trade fair building in the USA which was so complex that the designers were not convinced by their analyses.

They therefore decided to install a monitoring system to record the global movements of the roof. The first 2 years' records were used to establish a behaviour norm against which all future performance could be judged. A somewhat similar system was also used in the UK to assure the owners and the building control authorities that a building that appeared to be suspect when analysed using the normal building codes was indeed satisfactory.

This last example is perhaps the pointer to a greatly improved future for structural monitoring. Structural design is essentially an analytical process for which assessments (guesses) are made regarding the properties of the materials used for the construction, the accuracy and satisfactory construction of a building and also the environment in which it will have to function. No amount of general research can establish these data with complete accuracy. Therefore a complex system of factors (sometimes misnamed safety factors) exists to reflect our ignorance. The monitoring of actual structural performance whilst a building is in service can ensure that the assessment of safety is not solely based on prediction. This will allow the designers to relax their factors in the knowledge that there will be an early warning system to ensure the safety of the building users and to indicate the need for appropriate renovation/maintenance programmes and so enable the owners to keep their property in use for the maximum time.

References

Bate, S.C.C. (1974) Report on the failure of roof beams at the Sir John Cass's Foundation and Red Coat Church of England School, Stepney. *BRE Current Paper CP58/74*. Building Research Establishment, Watford.

Bate, S.C.C. (1984) High alumina cement concrete in existing building structures. *BRE Report*. Department of the Environment, Her Majesty's Stationery Office, London.

BRE (1973) *Report on the collapse of the roof of the assembly hall of the Camden School for Girls*. Her Majesty's Stationery Office, London.

BRE (1981) Carbonation of concrete made with dense natural aggregates. *BRE Information Paper 6/81*. Building Research Establishment, Watford.

BRE (1985a) The durability of steel in concrete. *BRE Digests 263, 264 and 265*. Building Research Establishment, Watford.

BRE (1985b) The structure of Ronan Point and other Taylor Woodrow–Anglian buildings. *BRE Report*. Building Research Establishment, Watford.

BRE (1986) Determination of the chloride and cement contents of hardened concrete. *BRE Information Paper IP21/86*. Building Research Establishment, Watford.

BRE (1987) The structural condition of prefabricated reinforced concrete houses designed before 1960. *Compendium of Reports, Ref. AP25*. Building Research Establishment, Watford.

BRE (1988) The structural condition of Easiform cavity-walled dwellings. *BRE Report BR130*. Building Research Establishment, Watford.

BRE (1989) The structural condition of Wimpey no-fines low-rise dwellings. *BRE Report BR 153*. Building Research Establishment, Watford.

Chambers, J.H. (1976) *Cyclical Maintenance for Historic Buildings*. US Department of the Interior, Washington DC.

Church of England (1980) *A Guide to Church Inspection and Repair*. Church Information Office, London.

CIRIA (1986) Structural renovation of traditional buildings. *CIRIA Report 111*. Construction Industry Research and Information Association, London.

Collacott, R.A. (1985) *Structural Integrity Monitoring*. Chapman & Hall, London.
Currie, R.J. (1986) Carbonation depths in structural quality concrete. *BRE Report BR 75*. Building Research Establishment, Watford.
DoE (1989) The effects of acid deposition on buildings and building materials. *Building Effects Review Group Report*. Department of the Environment, London.
Feilden, B. (1985) Towards a maintenance strategy. *Proceedings of a conference on Building Appraisal, Maintenance, and Preservation*. Department of Architecture and Building Engineering, University of Bath.
Gosselin, D., Simard, B. and Seckin, M. (1987) In Situ Testing to Failure of a Warren Truss Hangar. *Structural Assessment*, eds. Garas, F.K., Clarke, J.L. and Armer, G.S.T. Butterworths Scientific Limited, London.
Griffiths (1968) *Report of the Inquiry into the Collapse of Flats at Ronan Point, Canning Town*. Her Majesty's Stationery Office, London.
Ishizuka, Y. (1981) The degradation and prediction of service life of building components. *Proceedings of the 2nd International Conference on Durability of Building Materials and Components*. US Department of Commerce (National Bureau of Standards), Washington DC.
Mainstone, R.J. (1988) *Hagia Sophia*. Thames & Hudson, London.
Mitchell, T. (1989) PRC Houses—the SSHA Experience. *The Life of Structures*, eds. Armer, G.S.T., Clarke, J.L. and Garas, F.K. Butterworths Scientific Limited, London.
Sowden, A.M. (1985) Some aspects of bridge maintenance. *Proceedings of a conference on Building Appraisal, Maintenance and Preservation*. Department of Architecture and Building Engineering, University of Bath.

Index